EARTHQUAKES AND VOLCANOES

Readings from
**SCIENTIFIC
AMERICAN**

EARTHQUAKES
AND VOLCANOES

With Introductions by
Bruce A. Bolt
University of California, Berkeley

W. H. Freeman and Company
San Francisco

Cover photograph by Mark Hurd Aerial Surveys.

All of the SCIENTIFIC AMERICAN articles in *Earthquakes and Volcanoes* are available as separate Offprints. For a complete list of articles now available as Offprints, write to W. H. Freeman and Company, 660 Market Street, San Francisco, California 94104.

Library of Congress Cataloging in Publication Data

Main entry under title:

Earthquakes and volcanoes.

   Bibliography: p.
   Includes index.
   1. Earthquakes—Addresses, essays, lectures.
2. Volcanoes—Addresses, essays, lectures.
I. Bolt, Bruce A., 1930–   II. Scientific American.
QE521.E2   551.2   79–21684
ISBN 0–7167–1163–X
ISBN 0–7167–1164–8 pbk.

Printed in the United States of America

9 8 7 6 5 4 3 2

PREFACE

Earthquakes and volcanoes are awe-inspiring manifestations of the forces that shape our planet. The linking of these violent outbreaks by scientific thinkers can be traced back to the early Greeks, who witnessed earthquakes and eruptions along the shores of the Aegean and Adriatic Seas. In modern times, scientific studies of earthquakes and volcanoes have shown that connections between them are more remote than the early Greeks supposed. Earthquakes are not generally forerunners of volcanic activity, and the eruption of a volcano does not usually foreshadow a sharp earthquake. Yet, fundamental social and geological questions are common to the occurrence of both earthquakes and volcanoes, and it is still worthwhile to think about them together.

Our first and perhaps foremost concern is the hazard of seismic and volcanic activity to a growing world population. Millions of people live in earthquake country and in the shadow of active volcanoes. In the last 500 years, about 3 million people have died from earthquakes, and another 200,000 people have died from effects of the more than 500 active terrestrial volcanoes. Most of these deaths have been immediate, but deaths have also occurred from such secondary effects as starvation caused by destruction of food sources and tsunamis (tidal waves) produced by sudden sundering or collapse of the earth's crust. Yet, the widespread fatalism of past times has been replaced by the will to fight back against such disasters. We are now trying to control the rate of release of energy, to provide adequate warnings, and to construct buildings so as to minimize damage.

Today we are also seeking ways to use these natural sources of energy. Earthquakes are produced by the rapid release of strain energy stored up in the rocks of the earth's crust. The greatest earthquakes are produced by sudden slips along major geological faults, sometimes with offsets of tens of meters taking place in a few seconds. Already, in a modest way, small earthquakes provide clues to the location of geothermal fields, and analysis of seismic waves can provide information on the size of these valuable energy sources. For decades, steam and hot water from volcanic areas have been used to produce electricity and to heat buildings in a pollution-free way. Indeed, overall, the cost to society of volcanic eruptions is much outweighed by the long-term benefits of volcanoes as sources of energy and as sources of fertile lands, which give a bountiful life to millions of people who cluster around these monuments of lava and ash.

Study of earthquakes and volcanoes has also provided us with much fundamental information about the geology of the earth. Perhaps it is less apparent that other scientific disciplines have also benefitted from such research. For example, in engineering, the study of structures and soils has been vitalized by the need to consider the dynamic effects of earthquakes. The study of properties of matter at high pressures and temperatures inside the earth has affected developments in solid state physics, and the study of seismic waves has greatly stimulated general wave theory. Recordings of earthquake waves at the many hundreds of seismographic observatories around the world provide large quantities of numerical observations for advanced statistical analysis, and the evaluation of risk from both earthquakes and volcanoes is also the subject matter of statistics and engineering. There has always been an intimate connection between astronomy and seismology. Seismographs in early times were usually established at astronomical observatories, where accurate time was kept, because seismological data on properties of the earth and moon are useful for more quantitative studies of the solar system. Study of earthquakes has even affected politics. For over twenty years, attempts to get a comprehensive test ban treaty that would control the testing of underground nuclear weapons have involved the study of seismic waves from natural earthquakes and underground explosions.

For half a century, analyses of seismic waves produced by earthquakes have provided detailed information on the constitution of our planet and the physical properties of its deep interior. In the beginning, seismologists saw "through a glass, darkly," but nowadays improvements in instrumentation, planning of experiments, and theoretical procedures have enabled a clearer picture of fine structure to emerge. Of course, there are still sharp differences of opinion, and much research remains to be done. Studies of volcanoes are also providing information on the present and past states of the earth's interior. In particular, knowledge of the chemical properties of the rocks of the interior depend on chemical analysis of lava flows from volcanoes in various geological regions. Changes in magma composition with time are observed at many volcanoes; these can be related to the temperatures and pressures at which the magma originated, thereby yielding the history of the volcanic sequence over geologic time.

Studies of earthquakes and volcanoes play a crucial role in speculations on the energetics of the past history and future evolution of the earth. The properties of both earthquakes and volcanoes have been reexamined in recent years in the context of the new version of the theory of continental drift called *plate tectonics*. As explained in a number of articles in this book, this theory holds that the outer 80 km or so of the earth (the lithosphere) is divided into six major plates and about a dozen minor ones. These plates or caps move slowly relative to one another.

The plate tectonic model has cut a dash with scientists and the public alike since its inception and has scattered new ideas throughout modern geology. Perhaps its main contribution to date has been to provide a sense of order to the various patterns and rates of occurrence of earthquakes and volcanoes near the earth's surface. We can now trace patterns of activity back at least 150 million years, or to Jurassic time. The theory of plate tectonics has also allowed predictions to be made on the types and frequencies of earthquakes and volcanic activity that might occur in a specified region. Because the theory of plate tectonics has allowed geological explanations to encompass more than local historical observations, it has become a key to the management, mitigation, and forecasting of earthquakes and volcanic activity.

This collection of articles from SCIENTIFIC AMERICAN covers the main aspects of earthquakes and volcanoes mentioned. There is some repetition, but this diversity of explanation is not necessarily a drawback for either the general reader or student. Each of the three sections is preceded by an introductory section consisting of questions and answers drawn from the articles themselves. It is hoped that this form will fix the attention of the reader on key points and aid in establishing connections between articles. The questions range from the simple to the more complex. It should be stressed that there is a fundamental unity of both subject matter and methodology among the articles. It often turns out, for example, that what seems to be a new technique is only a more recent development of a previously used technique.

The first section discusses properties of earthquake sources, seismic waves, and the way these are measured by instruments. It includes a wide-ranging essay on prediction of earthquakes describing the search for geological clues that might permit specific forecasting of time and place. The second section contains three articles that give an up-to-date summary of the procedures seismologists use to explore both the structure and deformation of the earth's interior. These articles describe the types of inferences that may be drawn from seismograms. The third section discusses volcanoes as fascinating natural features in their own right and also as appurtenances to the global heat engine. No simple, universal explanation of volcanism is yet available, but we must start with the flow of heat from inside the earth and the ways heat is transferred from place to place. Two articles describe the consequences of assuming great convection currents in the earth's mantle. These currents drive the lithospheric plates from the oceanic ridges, where they are produced, to the island arcs and trenches, where they subduct to their ultimate destruction. Volcanism and seismicity are common elements of this dynamic process.

August 1979   *Bruce A. Bolt*

CONTENTS

Note on cross-references: References to articles included in this book are noted by the title of the article and the page on which it begins; references to articles that are available as Offprints, but are not included here, are noted by the article's title and Offprint number; references to articles published by SCIENTIFIC AMERICAN, but which are not available as Offprints, are noted by the title of the article and the month and year of its publication.

EARTHQUAKES AND VOLCANOES

I EARTHQUAKE PROPERTIES

INTRODUCTION

The following four articles discuss some of the basic topics in seismology. A helpful way to summarize the common features of these articles is to pose a series of questions and provide short answers. Somewhat different definitions, emphases, and interpretations are given by the various authors, so the answers may not always represent the exact position of each author. These differences may be taken as a measure of the state of the art in seismology over the past decade or so. Study of earthquakes has progressed so rapidly in this recent period that the data base of each article should also be taken into account. It is interesting to compare some tentative positions and speculations in each article with subsequent developments even a year or so later. The questions and answers are meant to represent the current state of knowledge.

Q: How are earthquakes produced?

A: The most common type of earthquake is called *tectonic,* because it is produced when rocks break suddenly in response to geological forces within the earth. Sometimes we define a *volcanic* earthquake as one that occurs in conjunction with volcanic activity, but the mechanism of these earthquakes is probably the same as for tectonic earthquakes. Sudden shaking of the ground is also produced by other, minor causes, such as the collapse of underground caverns and mines, landslides, and large explosions of chemicals or nuclear devices.

Q: What is the mechanism for tectonic earthquakes?

A: Because rocks are elastic, energy is stored in them during deformation by tectonic forces. When the strain builds to a level that exceeds the strength of a weak part of the earth's crust, such as along a geological fault, opposite sides of the fault suddenly slip, producing elastic waves that spread out into the earth. This theory was first propounded by H. F. Reid and is called the *elastic rebound theory* of earthquake genesis.

Q: What is the evidence for the elastic rebound theory?

A: Some of the first persuasive evidence came from field observations of slip along the San Andreas fault in the 1906 San Francisco earthquake. In this part of California, geodetic surveys during the previous fifty years had shown that sites remote from the San Andreas fault had moved several meters horizontally relative to each other. Many geodetic and seismological studies elsewhere since 1906 have verified the concept of fault rupture, with opposite sides of the fault springing back to an equilibrium position in a matter of seconds. This rebound produces both heat and seismic waves.

Q: Describe the model of an earthquake source that is most useful in understanding the production of earthquakes.

A: In an idealized model of an earthquake source, rupture of the fault begins at a point on the fault surface called the *earthquake focus* or *hypocenter*. The rupture surface then spreads along the fault plane at a certain velocity and finally stops, after it has spread across an area of average length L and average width W. The orientation of this fault plane is specified by its angle of strike (angle between north and the direction of the fault) and its angle of dip from the horizontal.

Q: What types of waves constitute an earthquake?

A: Seismic waves are of two main types: body waves and surface waves. Body waves travel through the body of the earth; the fastest are called primary (P) waves, which are compression waves capable of propagating through both solid and liquid rock. The slower body waves (secondary or S waves) are transverse shear waves that can travel through solid but not liquid rock. Surface waves travel around the surface of the earth with velocities that are generally slower than those of P and S waves. Two kinds of surface waves predominate: Rayleigh waves, which have motions in the vertical plane aligned in the direction of travel, and Love waves, which have only horizontal shear motions. At considerable distances from the earthquake source, P, S, Love, and Rayleigh waves separate from one another reasonably clearly, but near the source they may be mixed in complicated ways.

Q: How do variations in seismic surface waves allow the character of the rock below the surface to be explored?

A: Surface waves of longer wavelengths penetrate into deeper layers, where rocks have different transmission speeds, and thus arrive at the seismograph at different times. This phenomenon, known as *dispersion*, shows clearly a difference between the average crustal structure under continents and oceans.

Q: What are typical wave amplitudes and periods of seismic waves?

A: Both wave amplitudes and periods have a wide range of values, depending on the size of the earthquake and the distance of the shaking from the source. The amplitude of seismic waves can range from micrometers (millionths of a meter) to many centimeters. Similarly, near the source, earthquake waves may be of high enough frequency to be audible to the human ear (20 Hz or so). At the other end of the period scale, great earthquakes may produce resonance movements of the whole earth, with periodicities that range up to 53 minutes. Seismographs must be designed to record these wide ranges of amplitude and period.

Q: What is the principle of the seismograph?

A: Various types of instruments are designed to record the earthquake as a function of time. The most common instrument uses a form of pendulum suspended from a weak spring. Because the mass of the pendulum tends to remain fixed in space, measurements of the ground motion relative to the mass reveal the motion of the earth at the site. In another form, suspended masses can be used to measure the fluctuations in gravity caused by very long-period seismic waves or free earth oscillations. A different type of instrument, the strain seismometer, detects changes in distance between one end of a rigid rod and the other.

Q: What is the main problem faced in deciphering seismograms?

A: Because the earth has a variable and, in places, a complicated structure, seismic waves that are produced at a seismic source are distorted during their passage. Further distortion occurs from the response of the seismograph itself.

Thus, to disentangle the wiggles on the seismogram, the three factors—source properties, path properties, and instrumental properties—must be considered.

Q: What are the main parameters for describing the essential characteristics of an earthquake source?

A: The most common parameters are the location of the hypocenter and the *epicenter* (the point on the earth's surface immediately above the hypocenter) and the dimensions of the ruptured surface, $L \times W$. Also important are the orientation of the fault surface and the amount of slippage that has occurred along the fault. Some measures of the strength of shaking and the total energy released in the earthquake are needed as well.

Q: What is meant by *earthquake magnitude*?

A: The most common instrumental way of measuring the size of an earthquake is to use a magnitude scale first developed for local earthquakes in California by Charles F. Richter. The magnitude is calculated by taking the common logarithm (base ten) of the maximum amplitude of the actual ground motion and allowing for the distance to the source. If more specific measurements are needed, the wave type must be specified. Thus, the maximum amplitude of the P waves gives a P wave magnitude, and the peak amplitude of the surface waves will give a surface wave magnitude. Nowadays, a number of magnitude scales are in use.

Q: Is earthquake magnitude directly related to the energy released in the earthquake?

A: No. The relation is an empirical one only, and its validity depends a great deal on the frequency of the seismic waves being measured. Roughly, an increase in Richter magnitude of one unit entails a thirty-fold increase in the earthquake energy.

Q: Are there measures of size that have a better basis in physics?

A: Yes. Just as the motion of a dynamical system can be determined from the moments of the applied forces, so the strength of an earthquake can be specified in terms of the *seismic moment*. The seismic moment is the product of the rigidity of the rocks times the average slip along the fault times the area of the rupture ($L \times W$). Another useful physical parameter is the drop in stress across the fault during the rupture. This is also related to the amount of fault slippage and the elastic properties of the rock in contact along the fault surface.

Q: How can the geometry of the fault surface that ruptured to produce the earthquake be determined from the recorded P waves?

A: The most common method, devised by P. Byerly, is to compare the directions of first motions of the ground produced by the arrival of the P wave at seismographic stations. As the geometry of the elastic rebound along the fault changes, so does the pattern of compressions (ground up) and dilatations (ground down) carried by the seismic waves as they are radiated away from it. By measuring this pattern, the orientation of two orthogonal planes, one of which corresponds to the ruptured fault, can be inferred.

Q: In designing buildings, dams, bridges, and other structures, what are the main requirements of engineers regarding earthquake properties?

A: Engineers depend heavily on the predicted time histories of the accelerations, velocities, and displacements of the ground at the site of the structure. The peak expected acceleration is an important scaling factor, as is the expected duration of the strong ground motion. In design, engineers often use a peak frequency spectrum, which indicates how the structure will respond to the applied ground motion.

Q: How is strong ground motion measured?

A: Because sensitive seismographs easily go off scale, rugged instruments that record the high frequency accelerations of the ground during the passage of the seismic waves are usually employed. These are triggered by the first strong *P* wave and are capable of recording ground accelerations up to the acceleration of gravity (1 g) or even greater.

Q: What is the amplitude of peak ground acceleration in areas of structural damage?

A: In some earthquakes, near the rupturing fault, peak accelerations in both the vertical and horizontal direction in excess of that of gravity (1 g) have been recorded at higher frequencies. The usual range in large earthquakes on firm flat ground at frequencies below 8 Hz is between 0.3 g and 0.6 g. Peak accelerations are not, however, well correlated with earthquake magnitude.

Q: What is strong motion seismology?

A: This is the branch of seismology that considers the high amplitude motion of the ground near the source of an earthquake. Near the earthquake source, seismic waves have not sorted themselves out clearly into wave groups, and the effect of rupture over an extended area of the fault plane (rather than at a point) is dominant.

Q: What are the free oscillations of the earth?

A: When a large earthquake occurs, it releases sufficient energy to set the whole earth vibrating for many hours or even weeks. The vibrations resemble the oscillations of a violin string when it is plucked. Just as a violin string will have many modes and overtones making up a musical note, so the earth has many fundamental modes and overtones of motion, which together give the spectrum of free terrestrial oscillations.

Q: What types of free oscillations are there for an elastic, nearly spherical body like the earth?

A: There are two classes of oscillations. The simplest is called the *torsional* or *T* modes. In these modes, the particles of rock move only horizontally and not vertically. These oscillations only take place in solids, and so they do not affect the liquid core of the earth. The second class is called *spheroidal* or *S* modes. In these vibrations, particles of the earth move both horizontally and vertically, and there can be movement in both the solid and liquid parts of the earth. These modes also affect the earth's gravity field, so they can be detected both by pendulum seismographs and by gravimeters, which measure fluctuations in gravity.

Q: What is the scientific value of studying the free oscillations of the earth?

A: Just as by listening to the tones from a plucked violin string one can determine information about the string's length and tension, so measurement of free earth oscillations gives information on the structure and physical properties of the deep interior. Because the various modes of vibration involve particle motions at different depths in the earth, their periods reflect the properties of layers at various depths of the interior. The gravest mode of oscillation, $_0S_2$, called the *football mode*, has a free period of about 53 minutes. When the earth is shaking in this vibration, nearly all parts of it are moving, from the surface to the center.

Q: Can measurements of the free oscillations also provide information on the source of the earthquake energy?

A: Properties of the source, such as fault length and dip, affect the combinations of vibrations for a particular mode. Thus, we may imagine a kind of terrestrial spectroscopy, in which the frequency spectra near each mode give information on the earthquake source.

Q: Does the damping of these free oscillations warrant study?

A: Yes. The tone of a ringing bell dies away with time, depending on the damping properties of the metal. In the same way, various modes of oscillations of the earth take various lengths of time to attenuate. A measure of this energy attenuation, called Q, can be related to the damping properties of the rocks at various depths within the earth.

Q: What are the main types of earthquake prediction?

A: In most public discussions, earthquake prediction means the forecasting of a specific time and place of an earthquake of specified magnitude. These predictions might be either short term (a matter of days or weeks) or relatively long term (months or years). A less publicized type of earthquake prediction is the specification of the strong ground shaking that will occur at a particular site in an earthquake produced by a rupture of a nearby fault.

Q: What are the main advantages and disadvantages of earthquake prediction of time and place?

A: Long-term prediction could lead to higher priorities being given to remedial action to strengthen nonresistant structures in the threatened area and to increase public educational programs and emergency preparedness. Short-term prediction might lead to placing general emergency preparedness measures on full alert and to temporary actions to reduce danger in homes, factories, and hospitals. Possible disadvantages include economic losses from false alarms, slowing and shifts in industrial development, and changes in insurance rates.

Q: What are some promising earthquake precursors?

A: The primary one at present is rapid but measurable deformation of the target area. Regional crustal strain and uplift and tilting of the ground surface, measured by geodetic methods, are aspects of this precursor. Changes in the speed of seismic waves through the area are also possible viable indicators. Anomalous changes in water level and radon gas concentration in wells are also high on the list of precursors. A series of small earthquakes (foreshocks) sometimes (but not always) occurs before the main shock.

Q: What physical models have been advanced to guide and interpret field observations on earthquake precursors?

A: The basic model of precursory effects assumes the growth of microcracks in the rocks of the crust as the strain builds up just before an earthquake. The opening cracks increase the volume of the strained rocks (called *dilatancy*) of the area. In turn, this changes velocities of seismic waves and perhaps causes uplift and local tilting, foreshocks, increased radon escape, and changes in electrical resistivity of the rocks.

Q: Are the durations of the premonitory changes correlated with the size of the predicted earthquake?

A: A considerable body of evidence so indicates. For example, a magnitude 5 earthquake appears to be preceded by anomalous behavior for about three months, whereas a magnitude 7 earthquake would be preceded by anomalous behavior for about ten years. Unfortunately, the most damaging earthquakes, such as those in 1906, involve such long rupture length that the basis for these correlations is not at all clear. Great earthquakes, however, can often be forecast in a general way along extensive fault zones where there are gaps in the historical seismicity.

Q: What are the main Japanese and Chinese prediction activities?

A: A research program in Japan, under way since 1965, has achieved important results on earthquake mechanisms. Some areas have been designated as likely sites of damaging earthquakes in the next decade or so, but no short-term prediction of a large earthquake has yet been made. Long-term, closely spaced measurements of strain and tilt are in progress.

Because of hazardous housing conditions for many millions of Chinese people, the government has embarked on a vigorous program of earthquake prediction. Perhaps hundreds of thousands of people have been involved in a number of predictions, with a mixture of false alarms and successes. A notable success was the short-term forecast of the Liaoning (Haicheng) earthquake of 1975, when people were evacuated and lives saved. A notable failure was the unpredicted 1976 Tangshang earthquake near Peking, which is reported to have killed 650,000 people.

Q: How is earthquake risk in a country evaluated?

A: The risk from a geological event is usually given in terms of the odds of occurrence of the event within a given time interval. Seismic risk maps for the United States have now been produced that show the probability of occurrence in a fifty-year period of various levels of seismic ground acceleration and velocity at any location. These are based on the historical earthquake record and estimates of activity of major geological faults. In the final assessment of risk, the types of structures and the population density must be included, because an earthquake (or volcanic eruption) poses no risk in an uninhabited area.

Q: Can earthquake sources be controlled?

A: Some research on the modification of earthquakes by man has been done, based on the observation that injection of water in a fault zone reduces frictional resistance and is followed by a series of (usually small) earthquakes. The engineered modification of large earthquakes along major faults, however, is not yet feasible.

Q: Why does the San Andreas fault receive so much attention from seismologists?

A: The San Andreas fault is an easily accessible example of a boundary between one major plate (the North American) and another (the North Pacific). It has been the locus of great historical earthquakes, is well instrumented, and passes near major urban areas.

Q: How is the horizontal slippage of the two plates in contact along the San Andreas fault consistent mechanically with the complex region of thrusting in the transverse ranges of California?

A: One model postulates two sections of the San Andreas fault, the northern section having had relative slippage of more than twice that of the section south of the Tehachapi and transverse ranges. It is suggested that the nonalignment of the strike of these two fault sections has produced a large crustal compression in the vicinity of their junction, thus creating the transverse ranges.

Q: What is the scale of displacements along major faults, such as the San Andreas in California?

A: Careful field mapping of stratigraphy allows the matching and age dating of rock units that have moved with respect to each other along the fault. Such measurements suggest a movement on the average of about a centimeter a year along the San Andreas fault over the past 10 million years. This is less than the 3 cm per year obtained from present geodetic measurements along the fault.

Q: Can the evolution of the San Andreas fault system be explained by plate tectonics?

A: D. Anderson presents one reconstruction successfully worked out on this basis. However, in mathematical terms, the model only demonstrates the existence of solutions; because of the number of degrees of freedom in assigning block motions, no model can yet be adopted as unique.

Q: Because areas of California have been relatively free of earthquakes during this century, are they now immune from seismic risk?

A: No. The total historical record of California is about 200 years, and there is geological evidence on segments of the San Andreas fault of average times of approximately 160 years between large earthquakes. Given the many long active faults in California and the time elapsed since the 1906 earthquake, a large earthquake is very likely before the end of the century.

Q: What is the *intensity* of an earthquake?

A: A qualitative (and the oldest) way of measuring the intensity of an earthquake is to use the observed effects of the shaking on the ground, on people, and on structures. The Modified Mercalli intensity scale runs from I to XII. Correlations between intensities and recorded ground motions provide a way to assess the properties of earthquakes in former times.

Q: What was learned from studies of the San Fernando earthquake of 1971?

A: This earthquake occurred in a densely populated modern urban environment. Many strong motion accelerometers happened to be placed in the region. The records obtained were used to test theories that predict ground motions and to check on earthquake codes and building practices.

Q: Could the San Fernando earthquake of 1971 have been predicted?

A: Unfortunately, appropriate surveillance instruments were not in place in the area before the earthquake. The geological, geodetic, and seismological evidence that was available gave no reason to believe that the San Fernando area was any more or less likely than any other region in coastal California to experience a large earthquake.

The Motion of the Ground in Earthquakes

December 1977

The slippage along a fault that produces an earthquake radiates seismic waves. Exactly how these waves shake the ground bears on the design of buildings and other structures in earthquake zones

Last year half a million people were killed by an earthquake that devastated the Chinese industrial city of Tangshan. In the western U.S. over the years earthquakes have caused considerable damage, although the number of fatalities has been relatively small. The low casualty rate has been partly due to the fact that many of the major earthquakes occurred either in sparsely populated areas or fortuitously quite early in the morning, when most large office and public buildings are almost empty. Over the past few decades, however, many earthquake-prone regions of the western U.S. have become further urbanized. In them more large buildings and facilities such as dams have been constructed or are being planned. If such structures were to fail during a future earthquake, large numbers of people could be killed or injured.

Today the attention of many seismologists is being focused on ways to reduce the hazards of earthquakes by learning how to predict their consequences. To many people the term earthquake prediction probably suggests determining the time, place and magnitude of future earthquakes. Equally important is determining which of many ways the ground is likely to shake during the earthquake, how strong the shaking will be and how long it will last. Knowledge of the ground motion that can be expected during an earthquake can make it possible to design structures that do not need unnecessary and uneconomic levels of strength in order to survive being shaken.

In order to predict both the occurrence of an earthquake and the ground motion it will generate it is essential to understand the characteristics of the earthquake source. So far most of our understanding of earthquake sources has come from measurements made during actual earthquakes at seismological stations some distance from the source. Such measurements yield information about certain average properties of the earthquake source, for example the dimensions of the original disturbance and the overall movement involved in it. Average properties are useful in elucidating how seismic energy is released and how it is transmitted over large areas; they have also been invaluable in probing the structure and nature of the earth's interior and in assessing the likelihood of large earthquakes in certain regions. Such average properties, however, yield little information about the details of the ground shaking in areas immediately surrounding the earthquake source. It is this kind of specific information structural engineers require. For that reason a number of seismologists are now beginning to investigate the details of earthquake sources. This important subject, which might be called strong-motion seismology, is still in its infancy but should grow rapidly.

Historically our understanding of the cause of earthquakes is relatively new. By the middle of the 19th century it had been observed that the damage caused by many earthquakes was concentrated in a narrow zone, which suggested that earthquakes had a localized source. It was not until the San Francisco earthquake of 1906, however, that it was recognized that earthquakes were caused by slippage along a fault in the earth's crust. In a classic study conducted shortly after the earthquake Harry F. Reid of Johns Hopkins University discovered that for several hundred kilometers along the San Andreas fault fences and roads crossing the fault had been displaced by as much as six meters. Moreover, precise geodetic surveys conducted before and after the earthquake demonstrated that the rocks parallel to the fault had been strained and sheared. On the basis of such observations Reid proposed the elastic-rebound theory of earthquakes.

According to the elastic-rebound theory, rocks are elastic, and mechanical energy can be stored in them just as it is stored in a compressed spring. When the two blocks forming the opposite sides of the fault move by a small amount, the motion elastically strains the rocks near the fault. When the stress becomes larger than the frictional strength of the fault, the frictional bond fails at its weakest point. That point of initial rupture, called the hypocenter, may be near the surface or deep below it.

From the hypocenter the rupture rapidly propagates along the surface of the fault, causing the rocks on opposite sides of the fault to begin to slip past each other. A portion of the frictional stress the rocks had exerted on each other before the rupture is suddenly and violently released; the rocks along the fault rebound, or spring back, to an equilibrium position in a matter of seconds. The elastic energy stored in the rocks is released as heat generated by friction and as seismic waves. The seismic waves radiate from the hypocenter in all directions, producing the earthquake. The point on the surface of the earth above the hypocenter is the epicenter of the earthquake.

In some cases the rocks rebound not in a period of seconds but over an interval of minutes, days or even years. The seismic energy radiated at any one time is then quite small. This slow process is known as aseismic slip or creep. Why the seismic energy is released violently in some cases and not in others is not well understood.

Although the physical details of the elastic-rebound theory are still uncertain, the conceptual model of the faulting process fits well with the current hypotheses of plate tectonics. Most earthquakes are generated in zones where the huge plates of the lithosphere, which make up the outer layer of the earth's surface, are shearing past each other.

The concept of slip along a fault is at the heart of virtually all studies of earthquake sources. Indeed, the concept developed largely from investigations of earthquakes along the San Andreas fault. The San Andreas is a very long fault but not a deep one; earthquakes caused by its slippage are confined to about the upper 15 kilometers of the crust. Yet the study of this one shallow fault has led to a model that successfully explains the deformation of the ground and the radiation of seismic waves from all types of seismic sources, ranging from the shallowest slips to ruptures as deep as 700 kilometers along the advancing edge of a plate plunging below another plate.

The way the ground is deformed and the nature of the seismic waves that radiate during the earthquake provide basic information about the earthquake source: its dimensions, its shape and its orientation. The seismic waves have a wide range of period and amplitude. When a fault slips, the rupture process

itself generally lasts between a fraction of a second (for a minor earthquake) and five minutes (for a major one). The waves generated by the fault's slippage can have periods ranging from essentially infinity down to less than a tenth of a second. The seismic waves with the longest period correspond to the quasipermanent deformation of the ground around the fault. The waves with the shortest period actually fall into the low audible range. The waves with periods of about an hour have a frequency that coincides with the resonance frequency of the earth, and they cause the entire planet to ring like a giant bell.

The amplitude of the seismic waves can range from micrometers (millionths of a meter) to tens of meters. The amount by which the seismic waves deform the ground decreases with distance from the earthquake. In the great Chilean earthquake of 1960, for example, the total displacement of some points immediately adjacent to the fault ranged up to 20 meters. At Los Angeles, a quarter of the way around the world, the maximum displacement of the ground was about two millimeters.

Since seismic waves span such a broad spectrum of period and amplitude, many different kinds of instruments and experimental techniques are needed to capture all the information radiated by an earthquake source. Repeated geodetic surveys of the earth's surface can monitor deformations of the ground created by seismic waves with periods ranging from days to years. A variety of different seismographs have been designed to record seismic waves with periods ranging from an hour to a hundredth of a second. Some instruments are so sensitive that they can detect motions as minute as one micrometer, which they magnify tens of thousands of times in order to record them on paper. Other instruments are so rugged that they can withstand the jarring accelerations of the most violent earthquakes.

The record produced by the seismograph—a seismogram—holds a great deal of information; even with the aid of a computer, however, deciphering that information is neither simple nor straightforward. The waves recorded on a seismogram after passing through the earth can be thought of as violin music

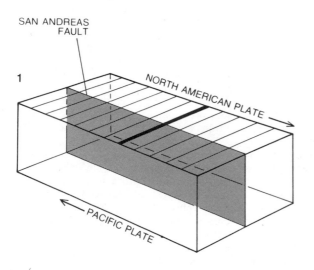

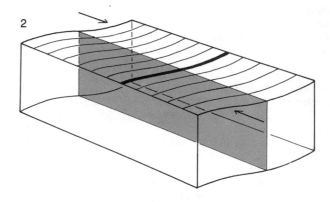

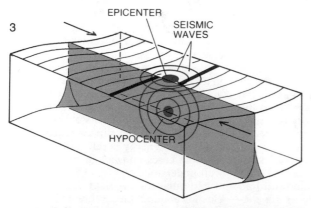

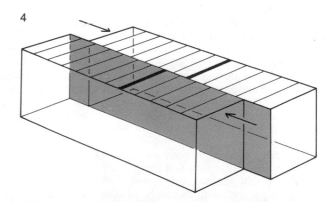

ELASTIC-REBOUND MODEL OF EARTHQUAKES assumes that two moving blocks of the earth's crust, each of which is part of a different tectonic plate in the earth's lithosphere, meet at a fault (*1*). Friction between the plates along the surface of the fault at first keeps them from slipping past each other, but the material around the fault is deformed by the stress (*2*). The deformation builds up until the frictional lock is ruptured at its weakest point, usually well below the surface (*3*). The rupture spreads out from that point, the hypocenter, radiating seismic waves as it does so. The point vertically above the hypocenter, where the seismic waves first reach the surface, is the epicenter of the earthquake. As the rupture spreads along the surface of the fault the blocks slip past each other, usually in a few seconds, coming to rest in a new equilibrium position (*4*). The stress around fault is relieved and ground rebounds to earlier state.

recorded on magnetic tape after first being transmitted over a telephone line that distorts the music. In this analogy the violin corresponds to the seismic source, the telephone line corresponds to the inhomogeneous elastic earth that distorts the signal passing through it and the tape recorder corresponds to the seismograph (which further distorts the signal as it is being recorded).

It is easy to correct for the distortion of the tape recorder. The challenge lies in trying to deduce something about the nature of the violin on the basis of the distorted sound received at the end of the telephone line. If one assumes that the telephone line is free of distortion, one might then reasonably conclude that a violin intrinsically produces a harsh sound. On the other hand, if one knows how a violin sounds when it is heard "live," one could use that knowledge to discover how the telephone line filters and distorts the music.

A similar problem faces the seismologist examining the record of an earthquake. In seismology the earth filter that distorts the seismic waves is complex because the internal structure of the earth is complex. As a result of decades of geological research, however, we now know much more about the earth's internal structure and how it distorts a seismic signal than we know about the earthquake source. Because the earthquake source is usually deep underground its seismic radiation cannot be "heard" firsthand. Seismologists must deduce the nature of the source by the indirect procedure of constructing a theoretical model of it, calculating the pattern of seismic radiation produced by the model, estimating how the seismic signal would be distorted as it propagated through the earth to the seismograph and comparing the synthetic seismogram with the actual seismogram recorded. By repeating the procedure several times with better information it is possible to refine the description of the earthquake source. Current models thus constructed attempt to describe the complex rupture process with relatively few parameters.

At the simplest level a model specifies the location of the hypocenter and the magnitude of the earthquake. At a more complex level the model includes the orientation of the fault surface underground and the direction of slip across the surface. The model can be made even more realistic by adding the dimensions of the entire area that ruptured, the average amount of slip across that area and the average length of time required for a point on the fault surface to be offset by the maximum amount. Since friction opposes the motion of the two sides of the fault past each other, it is believed that once a fault begins to slip, its direction cannot reverse. Such

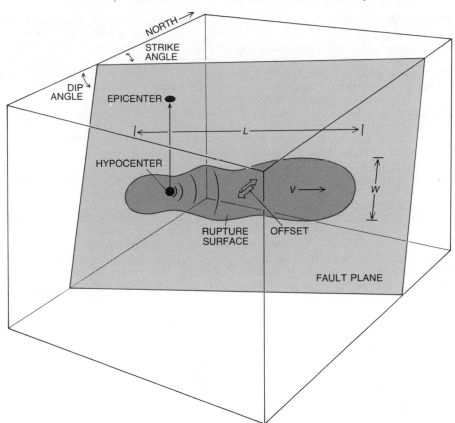

IDEALIZED MODEL OF EARTHQUAKE SOURCE suffices to describe most earthquakes with about a dozen variables. In the model the rupture begins at the hypocenter h kilometers below the surface, spreads across a fault plane at a velocity V and finally stops after growing into a region with an average length L and an average width W. The orientation of the fault plane is specified by its strike angle and dip angle. The slip between the two fault surfaces (*large arrows*) can have any orientation in the plane. On the average the slip requires τ seconds to reach its final offset. All these parameters are determined from recordings of the seismic waves.

models are quite successful in predicting the different types of seismic waves actually observed, particularly in predicting seismic waves with wavelengths at least as long as the dimensions of the fault.

The location of an earthquake can be determined by a procedure akin to triangulation, taking advantage of the fact that different types of seismic waves travel at different speeds. Seismic waves are of two general types: P waves and S waves. The P waves are longitudinal compression waves that travel through the deep interior of the earth, even propagating through the lower mantle and the liquid core. The S waves are transverse shear waves that travel through the solid portions of the earth.

P waves travel significantly faster than S waves. At a location close to the earthquake source the two types of waves will arrive fairly close together, but at one farther away the S wave will lag significantly behind the P wave. By observing the difference in arrival time between the two types of waves at any one station it is possible to calculate the distance of the earthquake from the station. Such a calculation from a single station does not determine the direction

of the earthquake, but when observations from three or more stations are combined, the precise location of the earthquake can be determined. If there are enough data, it is also possible to locate earthquakes from the P waves alone. In fact, this is the technique used by the National Earthquake Information Service in Golden, Colo., which collates earthquake data recorded all over the world and issues information about the position of an earthquake as soon as possible after each event.

The most widely recognized measure of the strength of an earthquake is the scale of magnitudes developed in the 1930's and 1940's by Charles F. Richter and Beno Gutenberg of the California Institute of Technology. The scale is based on the notion that ideally the magnitude determined should be an absolute measure of the energy released by the earthquake itself and should not be affected by the location of the seismographic station or the particular seismograph employed. The Richter method for determining the magnitude of an earthquake is quite simple. First, the seismologist measures the amplitude of the ground motion recorded in a certain specified part of the train of seismic waves. Second, he divides the recorded amplitude of the ground motion by the magnification of the particular seismograph to estimate the true ground motion at the seismographic station. Third, he calculates the common logarithm (the logarithm to base 10) of that ground motion. Fourth, he applies certain empirical corrections to that number to compensate both for the attenuation of the ground motion as it spreads out from the earthquake source and for the degree to which the response of the particular seismograph is influenced by local geological conditions.

The empirical corrections are applied so that for any given earthquake the same magnitude should be determined at all seismographic stations. In practice the magnitudes differ from one station to another, and an average magnitude is calculated from all of them. On the Richter magnitude scale larger numbers correspond to larger events. Since the scale is based on the common logarithm of the corrected ground displacement, each increase of one magnitude unit implies an increase of a factor of 10 in the amplitude of the ground motion. The magnitude scale is open-ended, and negative magnitudes have been measured.

Actually there are several magnitude scales in common use, each based on a different part of the seismic wave train. One is the scale of body-wave magnitude, measured from the P waves that travel through the body of the earth and reach the seismograph before any other waves. By convention, P waves with a period near one second are used in the magnitude determination. Another scale is the scale of surface-wave magnitude, measured from the dispersed waves that travel over the surface of the earth and reach the seismograph somewhat later. The surface waves employed have periods of 20 seconds. The two magnitude scales are cross-calibrated so that on the average both will yield the same magnitude when the earthquake being recorded has a magnitude of 6.75. By measuring the two magnitudes for a particular earthquake one obtains an estimate of the overall amount of seismic energy radiated in two quite different regions of the seismic spectrum. For a large earthquake the surface-wave magnitude is generally greater than the body-wave magnitude. This fact implies that the excitation in the long-period part of the spectrum increases faster with earthquake size than the excitation in the short-period part of the spectrum.

After the location and the magnitude of an earthquake have been determined from seismograms, the kind of information that can next be most readily obtained is the geometry of the earthquake source: the orientation of the fault in the earth, the dimensions of the portion of the fault plane that has slipped and the direction of the slip in the fault plane. Just as an array of radar antennas has a defined pattern of radiation, with large amounts of energy being beamed in some directions and small amounts in other directions, so also does an earthquake source have a defined pattern in which it radiates seismic energy. The radiation pattern not only determines the amplitude of the seismic signal in different directions but also determines how the seismic waves are polarized.

The radiation pattern can be understood by means of a simple experiment with a cube of foam rubber. Slit the top of the cube and push the two sides horizontally in opposite directions parallel to the slit. You will notice that the foam is compressed in two diametrically opposed quadrants and dilated in the other two quadrants. When a fault slips, the material around it is similarly compressed and dilated. The first waves emitted from an earthquake fault display the same distribution of compressions and dilations. The distribution of those waves on the surface thus reveals the orientation of the fault plane and the relative direction of the slip.

In the experiment with the cube of foam rubber, however, the quadrants of compression and dilation are clearly separated by two orthogonal lines; one line is the fault and the other line is perpendicular to the fault. Observations of the radiation pattern from an earthquake determine the orientation of two similar orthogonal planes, either one of which may be the earthquake fault. The ambiguity can be resolved if the orientation of the true fault plane is known

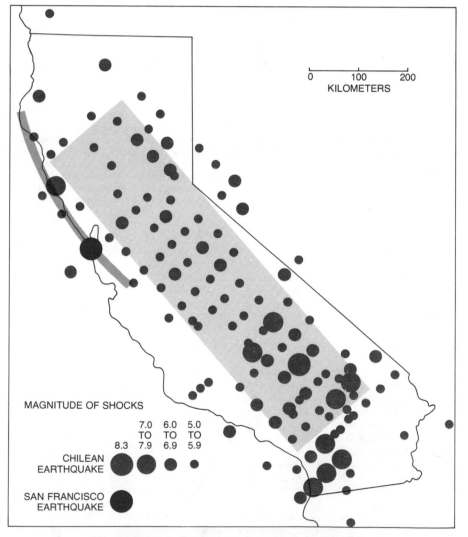

0   100   200
KILOMETERS

MAGNITUDE OF SHOCKS

| 7.0 TO 8.3 | 6.0 TO 7.9 | 5.0 TO 6.9 | TO 5.9 |

CHILEAN EARTHQUAKE

SAN FRANCISCO EARTHQUAKE

MAGNITUDE OF EARTHQUAKES is an inadequate measure of the actual size of large earthquakes. Both the San Francisco earthquake of 1906 and the Chilean earthquake of 1960 had a magnitude of 8.3. The area that ruptured in the San Francisco earthquake (*gray*), however, was approximately 15 kilometers deep and 400 kilometers long whereas the area that ruptured in the Chilean earthquake (*color*) extended to a depth equal to half the width of the state of California. The black dot represents the location of the epicenter of the San Francisco earthquake; the dots in color represent the locations of aftershocks of the Chilean earthquake with respect to its epicenter (*largest color dot*), superposed on the map of California for scale. The diameter of each dot represents the magnitude of each shock. Because earthquakes in California are caused by plates sliding past each other horizontally and not by plates subducting over each other as in Chile, no earthquake in California will be as great as earthquakes in Chile.

from the local geology. Alternatively, the orientation of the true fault plane can be determined from the pattern of aftershocks, smaller tremors that generally follow an earthquake, because the hypocenters of the aftershocks are usually scattered along the fault plane.

The information about the geometry of faults that has been amassed from earthquakes has been invaluable in developing the theory of plate tectonics. It has played a key role in identifying the faults between plates of the lithosphere and the relative motions of the plates. The seismic waves from an earthquake also yield information about the dimensions of the area that ruptured along the plane of the fault. The detail of the rupture area it is possible to resolve depends on the wavelength of the seismic radiation, just as in optics the wavelength of light limits the resolution of visual observation.

The area that ruptured in 1906, causing the magnitude-8.3 San Francisco earthquake, was 15 kilometers deep and 400 kilometers long; the area that ruptured in 1971, causing the magnitude-6.5 San Fernando earthquake in the Los Angeles area, was also 15 kilometers deep, but it was only 15 kilometers long. Seismic waves travel about four kilometers per second. The surface waves with a period of 20 seconds hence have a wavelength of some 80 kilometers. The 20-second waves might have provided a certain amount of detailed information about the source of the San Francisco earthquake, but with such waves the source of the smaller San Fernando earthquake would have appeared to be a point. By the same token, with seismic radiation having a period of several hundred seconds even the source of the San Francisco earthquake would have seemed to be a point.

Clearly a fault is not a point source. As a rupture propagates over the surface of the fault the point from which the seismic radiation is being emitted moves and causes the seismic waves emitted from one portion of the fault to destructively interfere with the waves emitted from another portion. The shorter the period, the more important the destructive interference. The period at which the interference first becomes noticeable can be used to estimate the dimensions of the fault. For example, the period might be about six seconds for a fault with dimensions of 10 kilometers by 10 kilometers or 60 seconds for one with dimensions of 100 kilometers by 100 kilometers.

In the 20th century about 55 earthquakes have been observed with surface-wave magnitudes ranging between 8.0 and 8.7, and no earthquakes have been observed with a surface-wave magnitude greater than 8.7. Actually two earthquakes near the upper end of the magnitude range may have the same

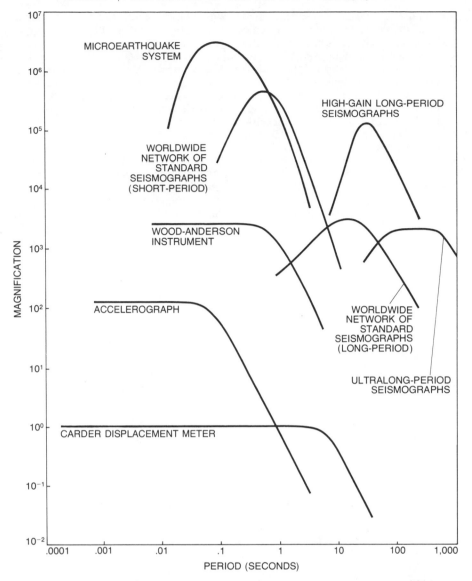

RESPONSE OF SEISMOGRAPHS of different types has been tailored to monitor seismic waves over a broad spectrum of period and amplitude. The magnification of the instrument is the number of times the instrument amplifies the ground motion so that it can be recorded. The amplitude of the ground motion in centimeters is approximately equal to the inverse of the magnification. For the most sensitive instruments the magnification is limited by ambient vibrations of the ground produced by wind and surf. The microearthquake system records small earthquakes within about 100 kilometers of the instrument. The Wood-Anderson instrument records moderate earthquakes at distances of several hundred kilometers. Moderate-sized earthquakes occurring almost anywhere in the world can be recorded on the short-period and the long-period systems of the Worldwide Network of Standard Seismographs or on special instruments such as the ultralong-period seismograph or high-gain long-period seismograph. Carder displacement meter and accelerograph record strong shaking close to fault.

surface-wave or body-wave magnitude and yet radiate vastly different amounts of seismic energy. In other words, for large earthquakes the magnitude scale becomes saturated.

The reason for this saturation is easily understood. The largest earthquakes rupture faults hundreds of kilometers long. If a fault is very long, it takes more time for a wave emitted from the farther end of the fault to reach the seismograph than it does for a wave emitted from the nearer end of the fault. Since the wavelength of a surface wave can be much shorter than the length of a very long fault, the part of the wave train from which the earthquake's magnitude

is measured will be emitted from only a fraction of the fault's area rather than from the entire fault. The result is that the strength of the earthquake appears to be less than it actually is, and the magnitude scale cannot accurately measure very large earthquakes.

A new measure of the strength of an earthquake, known as seismic moment, has recently come to the fore. Seismic moment is not as easy to measure as seismic magnitude, but it is a more physical measure of the size of an earthquake source. The seismic moment is determined by the Fourier analysis of seismic waves of such long period that the details of the rupture are smoothed out and the entire fault appears to be a point source. (The periods at which the seismic moment is determined increase with the size of the fault.) If the fault is "viewed" by such long-period waves, the slip from the unruptured state to the ruptured one appears to be instantaneous. The actual pattern of the seismic radiation emitted by the instantaneous rupture is mathematically equivalent to the theoretical pattern of radiation emitted by a model consisting of two hypothetical torque couples embedded in an unruptured elastic medium.

Each of the two torque couples can be visualized as a pair of small spheres,

with a thin wire attached to each sphere. The wires are pulled with equal force in such a way that one pair of spheres rotates in one direction while the other pair rotates in the opposite direction. The magnitude of the rotary force—the torque—exerted by each pair of spheres on the elastic medium is the moment. Since the two torque couples rotate in opposite directions, however, no net torque is applied to the medium. The two torques nonetheless deform the medium, radiating elastic waves in a characteristic pattern: a pattern identical with the one in which an earthquake source radiates seismic waves. From this model the moment of the seismic

radiation emitted by earthquakes can be calculated. The model has been named the double-couple source model.

The seismic moment measures the seismic energy emitted from the entire fault and not from just a portion of the fault, so that it is a fundamental measure of the magnitude of an earthquake. Hiroo Kanamori of Cal Tech has developed a new magnitude scale based on the seismic moment. The new scale extends the standard Richter scale so that it can accurately measure the strongest earthquakes without becoming saturated. For example, both the San Francisco earthquake of 1906 and the Alaskan earthquake of 1964 had a surface-wave magnitude of 8.3, but the seismic moment of the Alaskan earthquake was 100 times greater than that of the San Francisco one. On Kanamori's scale the magnitude of the San Francisco earthquake has been demoted to 7.9 and that of the Alaskan earthquake has been advanced to 9.2. The strongest earthquake on record is the Chilean earthquake of 1960, with a surface-wave magnitude of 8.3 and a seismic-moment magnitude of 9.5.

Seismic moment is more than just a convenient scale by which to rank earthquakes according to their magnitude. In 1966 Keiiti Aki of the Massachusetts Institute of Technology showed that the seismic moment is equal to the product of three factors: the average slip of the fault, the area of the rupture and the rigidity of the material that is faulted. Thus if one has independent measurements of the area of the rupture and the rigidity of the material, one can determine the average slip of the fault. The correlation between the average slip of a fault and the average strength of the resulting earthquake provides useful criteria for designing structures such as highways and pipelines that must cross active fault zones.

The total amount of slip accumulated from a number of earthquakes over a period of time also enables one to estimate the velocity at which the tectonic plates bounding the fault are moving past each other. By comparing that velocity with the velocity computed from independent geological, magnetic and geodetic evidence, it is possible to determine how much of the relative motion of the plates gives rise to earthquakes and how much gives rise to aseismic creep. It seems that in some areas, for example Chile, all the motion between plates is accomplished by earthquake slippage, and that in other regions, for example the Marianas arc in the western Pacific, the motion is accomplished by long-term steady creep.

The seismic moment and the dimensions of the fault also yield information about the amount of stress across the fault that is released during the earthquake. The drop in stress is only weakly

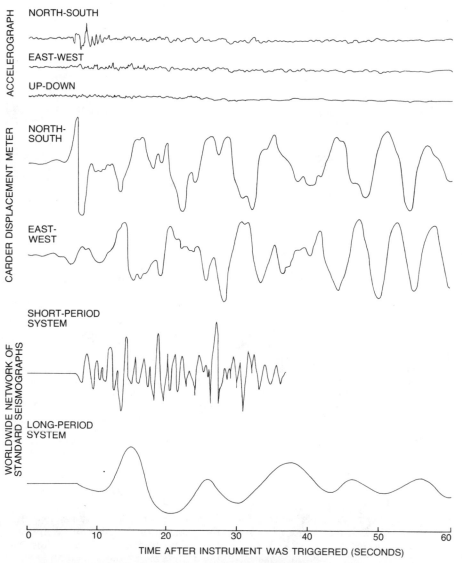

TYPICAL SEISMOGRAMS recorded by different instruments at the same site during the same earthquake can be remarkably different. The top two sets of curves are the recordings of an accelerograph and a Carder displacement meter at El Centro, Calif., from an earthquake at Borrego Mountain, some 60 kilometers away. Both instruments were triggered by the initial P wave, or compression wave, from the earthquake; the first strong pulse on each recording is the slower-traveling S wave, or shear wave, which arrived seconds later. The prominent reverberations on the recording from the Carder displacement meter are resonances of the seismic waves in the thick blanket of sediments in the Imperial Valley. The bottom pair of curves is the recording made at La Paz in Bolivia of the vertical component of the initial P wave from the same earthquake that was recorded by a short-period seismograph and a long-period seismograph in the Worldwide Network. By the time seismic waves had traveled to La Paz, a fifth of the way around the world, S waves (not shown) arrived approximately nine minutes later than P waves.

HORIZONTAL DISPLACEMENT of the ground during an earthquake in the Imperial Valley of California disrupted the regular pattern of trees in citrus groves. In this aerial photograph of an orchard seven miles east of Calexico, made shortly after the earthquake in 1940, the path of the San Andreas fault can be clearly traced diagonally across the groves west of the Alamo River. North is to the right.

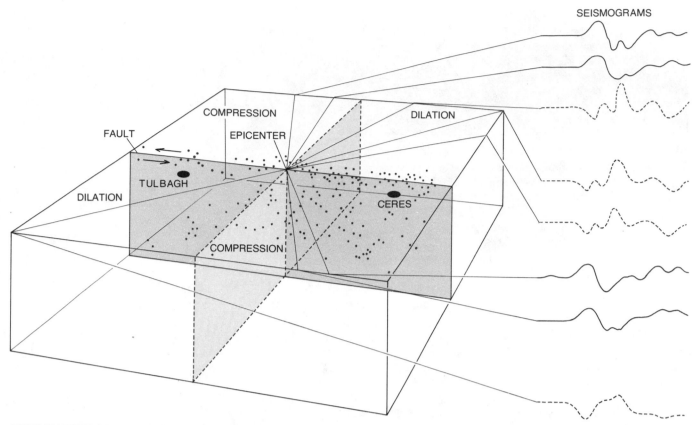

ORIENTATION OF A FAULT below the surface can be detected from the way the ground is initially compressed and dilated around the epicenter of an earthquake. This pattern of compressions and dilations is preserved in the seismic waves that are radiated by the earthquake source. In the illustration portions of seismograms (*right*) recorded during an earthquake near Ceres in South Africa show how the phase of the initial seismic waves received was shifted with azimuth between the source and the recording station. From this information alone the fault could be either of two orthogonal planes: the actual fault (*dark color*) or an imaginary plane perpendicular to it (*light color*). The path of actual fault can be determined from the location of earthquake aftershocks (*dots*) which lie along a single plane.

dependent on the magnitude of the earthquake. Most measurements during large earthquakes indicate that the drop in stress is between 10 and 100 bars. (A bar is 15 pounds per square inch.) The absolute, or total, stress on the faulted material could be considerably higher, but the radiated seismic waves are influenced only by the change in the stress across the fault and not by the absolute stress. Why the drop in stress should be essentially constant for earthquakes spanning such a great range of magnitude is under active debate; the explanation probably lies in the physical properties of the materials within the fault zone and in the forces driving the lithospheric plates.

The properties of the earthquake sources I have discussed so far have been deduced from seismograms made

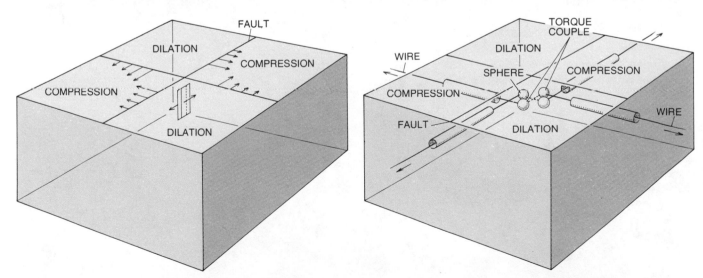

DOUBLE-COUPLE SOURCE MODEL is mathematically equivalent to the slippage of an earthquake fault. When a small fault slips (*left*), the material closest to it slips more (*longer arrows*) than the material farther away (*shorter arrows*). Thus the material around the fault is compressed and dilated. The same deformation pattern can also be obtained if opposite torques are exerted on two torque couples embedded in an elastic medium (*right*). A torque couple can be visualized as a pair of spheres with a wire attached to each sphere running through a frictionless tube to exterior of medium. When wires are pulled with equal force, elastic medium is deformed in same way as material around a fault. Moment, or amount of torque exerted, is a good measure of strength of earthquake producing the deformation.

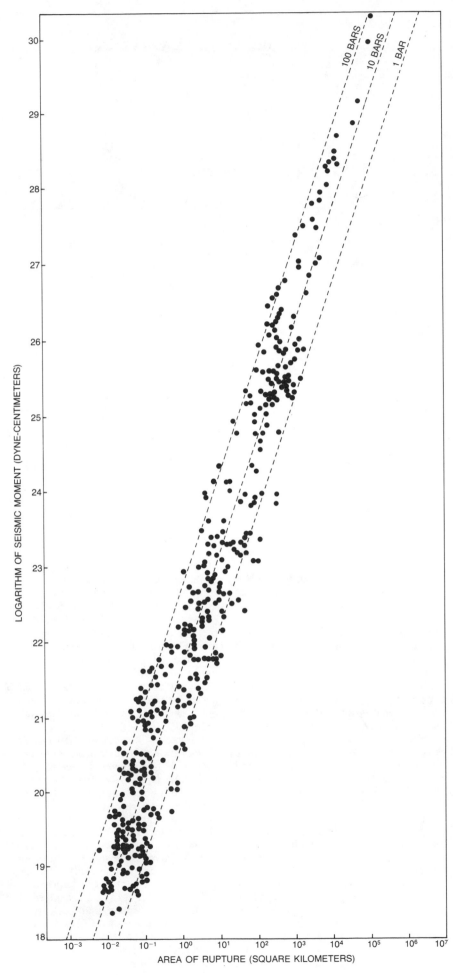

at stations far from the source. Observations at a distance, however, rarely make it possible to resolve the detailed structure of the source. Recently seismologists have been devoting an increasing effort to gaining an understanding of the intricate effects involved in the propagation of the rupture along the fault and in the distortion of the seismic radiation by geological heterogeneities near the fault. This understanding is essential for the design of structures to withstand ground shaking.

Man-made structures are particularly susceptible to earthquakes because the seismic waves have frequencies that coincide with the resonant frequencies of the structures (which range from a tenth of a hertz for large structures such as the Empire State Building up to 30 hertz or even higher for small structures such as systems of pipes in an industrial plant) and because the largest ground motions are usually in the horizontal plane. All buildings are inherently capable of withstanding large vertical forces (at least 1 g, or the force exerted on them by the earth's gravity) but special precautions must be followed in earthquake country to ensure adequate resistance to large horizontal forces.

In general the most destructive ground motions have wavelengths smaller than the dimensions of the earthquake fault. Therefore the ground motions are strongly influenced by the details of the rupture process, such as the speed at which the rupture travels over the fault surface, the frictional strength of the fault and the drop in stress across the fault. Geological heterogeneities in the path of the seismic waves can also affect the waves' amplitude and frequency; a seismogram recorded at two stations close to each other may differ significantly. In the past seismologists have rarely been lucky enough to have a good distribution of seismographs close to the source of a major earthquake, and the few seismographs that have been close to the fault have usually been shaken so violently that the recording pen was thrown off the paper. Accordingly the short-period seismic waves are not as well understood as the long-period ones.

In recent years several types of inex-

DROP IN STRESS across a fault during a large earthquake seems to be independent of the strength of the earthquake. The dots represent measurements of seismic moment obtained during many earthquakes with respect to the size of the rupture in square kilometers. Stress drop is inferred from measurements. Lines of constant stress drop are shown. Scatter in measurements for smaller earthquakes may be due in part to experimental error. A bar is a unit of pressure equal to 15 pounds per square inch; a dyne is a unit of force required to impart an acceleration of one centimeter per second per second to a mass of one gram.

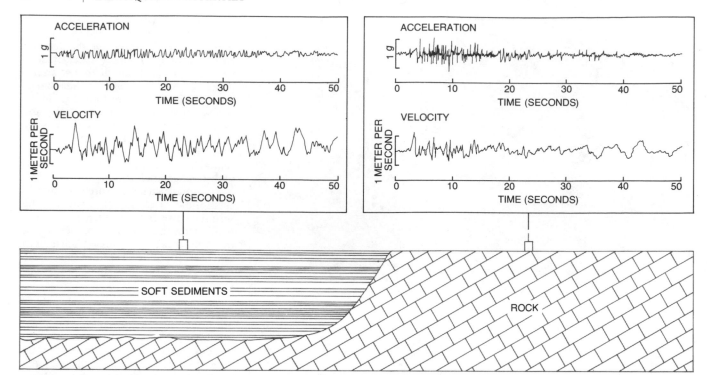

LOCAL GEOLOGY AFFECTS GROUND MOTION near the recording site. The waves propagating from the hypocenter up to the earth's surface slow down as they encounter the deformable rocks near the surface, and in general their amplitude increases in much the same way that the amplitude of an ocean wave increases as it approaches the shore. When soft sediments are subjected to strong shaking, however, the amplitude of the motion can actually be reduced. Seismograms at the right are hypothetical recordings of acceleration and velocity of ground for an area underlain by hard rock. Seismograms for a nearby area underlain by sediments (*left*) show that the ground moves faster but amplitude of its acceleration is less. Acceleration is given in terms of *g*, acceleration of gravity at earth's surface.

pensive, rugged and reliable low-magnification instruments have been designed and installed in large numbers near many earthquake faults. The most widely used instrument is the accelerograph, which measures the acceleration of the shaking ground. There are now more than 1,200 accelerographs on station in California alone. Even with so many instruments now in operation we still do not know much about the ground motions close to a fault during a severe earthquake. So far only two useful recordings of an earthquake of magnitude 7.0 or greater have been obtained within 40 kilometers of a fault, and one of them was obtained during an earthquake in the U.S.S.R. To a large extent this lack of data is due to the fact that there have been no large earthquakes in the U.S. in the four or five years since most of the accelerographs were installed.

The measurements that do exist have been the main resource for estimating the strength of the ground motion. The few recordings close to faults have had a disproportionate influence on earthquake engineering design, even though these data may not be truly representative of the motions close to future earthquakes. As might be expected, the few close-in recordings have received intensive scrutiny. For example, an accelerograph on a rock abutment near the Pacoima Dam in California during the San Fernando earthquake recorded a peak acceleration of nearly 1.5 *g*, the largest

acceleration yet recorded near an earthquake. The record was obtained in a region of exceptionally rugged cliffs and hills, and numerical simulations of the propagation of the seismic waves suggest that the topography may have amplified the ground acceleration by as much as 50 percent with respect to the motions that would be expected on flat ground.

At distances beyond 10 or 20 kilometers from the fault there are a fair number of recordings for earthquakes of magnitude less than 7.0. It is convenient to study the peak acceleration of the ground, expressed in terms of the acceleration of gravity at the earth's surface (*g*), which can be measured directly from the accelerograph records. The peak acceleration expected is widely used by engineers to specify the ground motion a structure should be able to withstand. The peak acceleration of the ground decreases with distance from the fault, both because the seismic waves spread out as they propagate away from the source and because their energy is attenuated by the slight inelasticity of the rocks through which they propagate. Between 20 and 200 kilometers from the fault the peak acceleration decreases approximately as the inverse square of the distance from the fault.

The data that have been obtained 20 kilometers or more from the fault imply that the peak acceleration of the ground

is correlated with the earthquake's magnitude. Contrary to what one might expect from the definition of magnitude, however, earthquakes differing by one unit of magnitude do not generate peak accelerations differing by a factor of 10. Moreover, the few available data obtained some 10 kilometers from the fault indicate that very close to an earthquake, peak acceleration is hardly correlated with magnitude at all. For example, an accelerograph close to a fault near Oroville, Calif., recorded a peak acceleration of .6 *g* during an earthquake of magnitude 3.4 but another instrument near a fault in the Imperial Valley recorded a peak acceleration of only .4 *g* during an earthquake of magnitude 7.1.

The lack of correlation between peak acceleration and magnitude is easily understood. The seismic waves measured by accelerographs have a dominant frequency of about four hertz, much higher than the frequency at which the magnitude of the earthquake is measured. For all earthquakes but the smallest, seismic waves with a frequency of four hertz have a wavelength much shorter than the dimensions of the fault. Thus peak acceleration is not a good measure of the strength of large earthquakes, and for the same reason that magnitude is not. The duration of the ground motion is probably much better correlated with earthquake strength.

Strong-motion seismology is a new

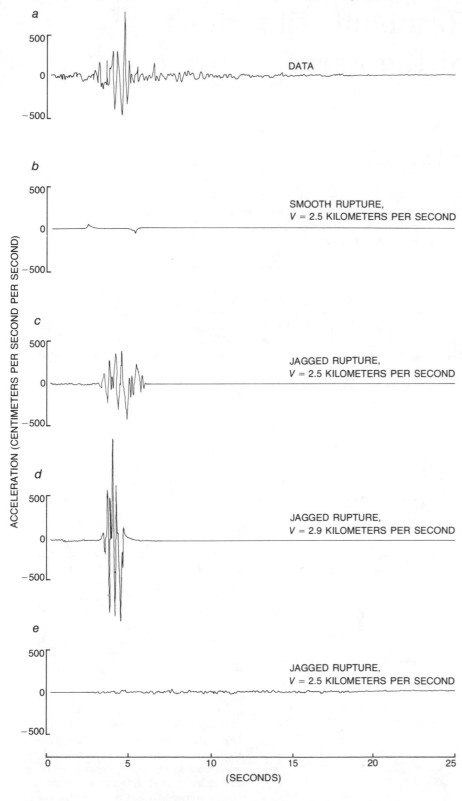

SYNTHETIC ACCELEROGRAMS were constructed on a computer by the author and William B. Joyner of the U.S. Geological Survey in order to determine experimentally how an earthquake generated observed ground shaking. An actual accelerogram is shown at the top (*a*). If the earthquake were produced by a smooth rupture of the fault propagating toward the theoretical seismographic station, its accelerogram would consist of a few simple isolated peaks corresponding to the radiation emitted as the rupture started and stopped (*b*). The peaks are small because the rupture was constrained to have a gradual acceleration and deceleration at the ends of the fault. Actual data, however, generally show a more continuous shaking. To simulate this shaking random fluctuations were added to the amount by which the fault slipped. The resulting theoretical curve looked more like the actual data (*c*). Next the author let the rupture propagate toward the theoretical accelerograph at a velocity close to the velocity of seismic waves in the surrounding material. The seismic radiation from the fault then arrived in a sharp peak (*d*). When rupture propagated away from theoretical accelerograph, however, the seismic radiation was spread over a longer time interval and its amplitude was reduced (*e*).

disdiscipline, and there are many unknowns in it. In the future data obtained by means of accelerographs and other instruments close to earthquake faults should provide information about both the complexities of earthquake sources and the ground motions they generate. The theoretical and computational models of the seismic source should also improve. Such information will be of direct value to engineers designing major structures. Until that information is available, however, architects and engineers must continue to design buildings on the basis of the few data that do exist, some simple theoretical scaling arguments and plain educated guesses.

Seismologists studying models of the strong ground motions near faults are just beginning to recognize that many of the problems facing them have an essentially statistical character. For years engineers designing major structures in earthquake zones have treated accelerograms of short-period motions as recordings of random noise. On that basis they have devised many ways to generate random series of short-period motions that look much like the accelerograph recordings. The random series were generated in such a way that they matched certain constraints derived from existing data, but they paid scant attention to the physics of the earthquake source. This engineering approach is certainly a reasonable first approximation on which to base the design of a building, but it is of little value in determining what is actually happening below the ground.

Seismologists, on the other hand, have tried to predict the ground motion from earthquakes purely on the basis of deterministic models. In these models earthquake sources have been idealized as simple faults in layer-cake geological structures. Such deterministic models have been relatively successful in predicting only the long-period components of the ground motions.

Clearly the time has come to merge the engineer's statistical view with the seismologist's deterministic one. A number of seismologists are now attempting such a synthesis. Predictions of the ground motion are, however, only as good as the statistical distributions incorporated into the model and physical knowledge of the earthquake source: the properties of the fault surface and of the surrounding rocks and soil. For that information we must not only study existing strong-motion recordings but also draw on other fields such as rock and soil mechanics. I foresee an exciting future in which the skills and the learning of many disciplines, ranging from classical seismology to soil engineering, are combined to gain a better understanding of the nature of earthquakes and to reduce the hazards they create.

Resonant Vibrations of the Earth

by Frank Press
November 1965

When a major earth quake occurs, the entire earth vibrates like a ringing bell. These extremely slow "free oscillations" yield information on the structure of the earth's crust and mantle

Pluck a taut string and it vibrates; strike a bell or a gong and it vibrates. The vibrations are called free oscillations, and they are excited in any mechanical system that is disturbed from equilibrium and then left alone. The earth also vibrates when it is disturbed: an earthquake sets the entire globe to oscillating like a bell for weeks or months. In the past five years geophysicists have learned how to detect these global free oscillations and analyze them to obtain information on the structure of the earth.

The value of free oscillations to the geophysicist arises from the fact that the total vibration of a body can be regarded as a superposition of independent harmonic motions called normal modes. Each mode can be excited by itself, but complex vibrations involve the superposition of large numbers of modes. The characteristic frequencies of these modes are determined solely by the nature of the vibrating body—not by the special circumstances that excite the vibration.

In the case of the earth all possible elastic motions of the planet can be represented by a superposition of free oscillations. Even the complex jumble of vibrations that emanates from an earthquake, which one usually thinks of as a series of different kinds of propagating waves, can be so represented. This follows from a dual aspect of waves: when a propagating wave interferes with itself, it is manifested as a "standing" wave. For example, the motion of a plucked violin string can be completely represented either by (1) a traveling wave that leaves the point at which the string is plucked and is reflected back and forth between the ends of the string or by (2) the modes, or standing waves, that are characteristic of the string [*see illustration at left*]. The wave motion by which seismic energy is transmitted is conventionally considered as a traveling disturbance affecting one part of the earth at a time; it can be considered just as correctly, and from a more general point of view, as belonging to some mode of vibration of the earth as a whole. This approach yields two dividends. It provides information on the structure and other characteristics of the earth as a whole, and it provides a "handle" with which to get at the information in certain very long seismic waves that are of special interest to geophysicists but that are

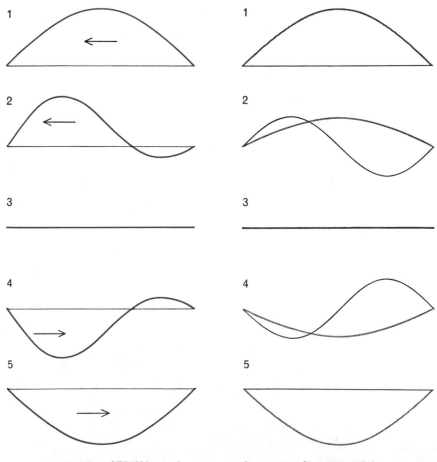

VIBRATION OF A STRING can be represented as a traveling wave (*left*) or as two standing waves (*right*). In this case the traveling wave is the result of superposition of the first two modes of vibration of the string, vibrating 90 degrees out of phase. Any wave motion, even the most complex, can be represented as the sum of a large number of modes.

FIRST FOUR MODES of a vibrating string, the fundamental and first, second and third overtones, are illustrated. The string vibrates between the dark- and light-colored positions. The mode number is one less than the number of nodes, or points of zero displacement.

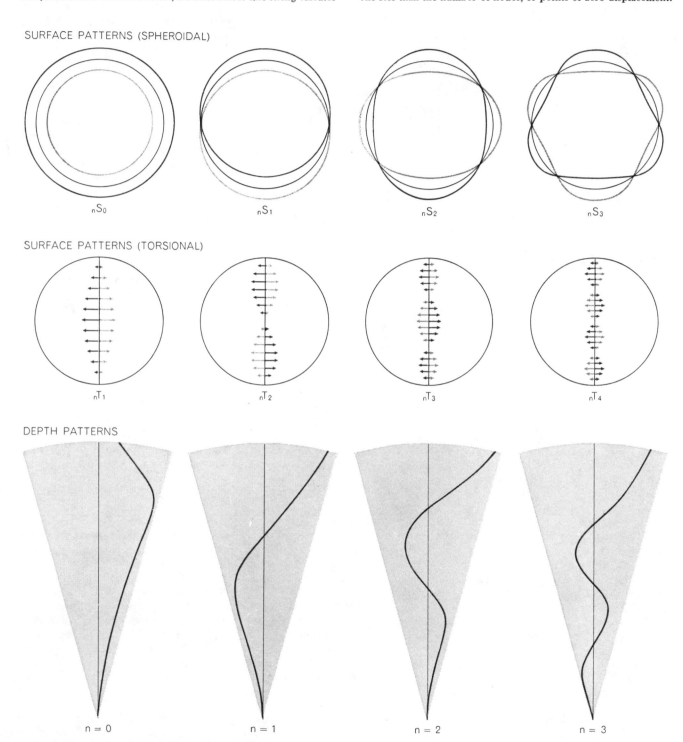

SURFACE PATTERNS (SPHEROIDAL)

$_nS_0$ $_nS_1$ $_nS_2$ $_nS_3$

SURFACE PATTERNS (TORSIONAL)

$_nT_1$ $_nT_2$ $_nT_3$ $_nT_4$

DEPTH PATTERNS

$n = 0$ $n = 1$ $n = 2$ $n = 3$

MODES OF A SPHERE are more complex. Here a few of the surface and depth displacement patterns are shown. In spheroidal (S) modes there is both radial and horizontal vibration; in torsional (T) modes, particles vibrate on a spherical surface. The subscript at the right of the designation of each surface pattern describes the pattern in terms of variation in latitude; in these examples the vibrations are assumed to be independent of longitude. The subscript n gives the depth pattern, or number of nodal surfaces. Any spheroidal or torsional mode can combine with any depth pattern; that is, it can be repeated at any number of levels.

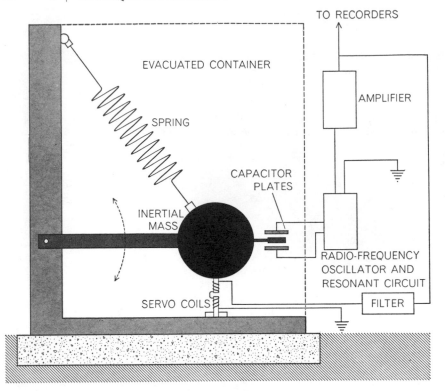

TO RECORDERS

EVACUATED CONTAINER

AMPLIFIER

SPRING

CAPACITOR
PLATES

INERTIAL
MASS

RADIO-FREQUENCY
OSCILLATOR AND
RESONANT CIRCUIT

SERVO COILS

FILTER

VERTICAL PENDULUM SEISMOGRAPH designed by Ralph Gilman of the California Institute of Technology detects the vertical component of long earthquake waves. The mass tends to remain stationary as the earth moves. Relative motion at the capacitor plates generates an electrical signal that is fed to an analogue and a digital recorder. The filter feeds back spurious signals, representing motions other than those of the desired waves, to coils that keep the mass centered. The base of the instrument is two feet long.

particularly difficult to define and analyze as individual waves.

The normal modes of a sphere can be grouped in two classes: "spheroidal," or S, modes and "torsional," or T, modes. In the T modes each particle of a body vibrates back and forth on a spherical surface—on the outside of the body or on some interior "shell." The S modes involve not only this horizontal motion but also a radial motion in and out from the center. Now, in the case of a vibrating string only one integer is required to describe a mode of vibration: the "mode number" is one less than the number of nodes, or points of zero displacement [see top illustration on preceding page]. It takes two integers to describe the modes of a vibrating membrane such as a drumhead. In the case of a sphere, which has three dimensions, three integers are required; they are designated l, m and n.

The integers l and m define the pattern of surface displacements, or of lines of nodes, with regard to the source of the disturbance. One integer, l, fixes the variation with latitude: the angular distance (measured at the center of the sphere) from the source. The other, m,

fixes the variation with longitude, or azimuth. When the oscillations are independent of longitude, there are l lines of nodal latitude in the surface pattern for spheroidal modes and $l - 1$ lines for torsional modes. The index n specifies the pattern of displacements with depth: the number of nodal surfaces [see bottom illustration on preceding page].

A mode of vibration can be completely described by either of the following two symbolic expressions: $_nS_l{}^m$ or $_nT_l{}^m$. A double infinity of modes and frequencies is possible. There are an infinite number of surface patterns as l varies from 0 to infinity and, for each l, there are a fundamental mode $(n=0)$ and an infinite number of overtones as n varies. In a nonrotating sphere the frequencies are independent of m. In the illustration (in which m is held constant so as not to affect the patterns) it can be seen that the sphere expands and contracts radially in the $_nS_0$ mode, "shakes" in the $_nS_1$ mode and becomes football-shaped in the $_nS_2$ mode. In mode $_nT_2$ the hemispheres rotate in opposite directions around the same axis; in higher T modes the direction of rotation alternates with smaller changes in latitude.

The theory of elastic vibrations in

spheres has been under development for a long time; it was discussed by the French mathematician Siméon Denis Poisson in 1829 and was worked out in detail and applied to the earth by several noted British mathematical physicists at the end of the 19th century and the beginning of this century; more than 50 years ago A. E. H. Love even made an estimate of one hour for the period of the earth's mode $_0S_2$, assuming a homogeneous earth with the mean elastic properties of steel. It was only recently, however, that a fortuitous combination of circumstances led to the observation and theoretical verification of the earth's free oscillations. Hugo Benioff at the California Institute of Technology, Lucien LaCoste of LaCoste & Romberg and Maurice W. Ewing and I at Columbia University's Lamont Geological Observatory developed instruments sensitive to long-period earth movements and these instruments were deployed around the world. Electronic digital computers became available, along with new techniques for the interpretation and numerical spectral analysis of seismic waves. And Chaim Pekeris and his colleagues at the Weizmann Institute of Science in Israel made advances in the theory of wave propagation, particularly for models of the real earth.

The major effort in instrumentation has been to improve the response to long-period, low-frequency waves [see "Long Earthquake Waves," by Jack Oliver; SCIENTIFIC AMERICAN, March, 1959]. This has required the development of electronic transducers that are selectively sensitive to very slow earth movements, with filters that minimize background noise and maximize the signal. Transducers that sense small changes in capacitance detect the movements of the inertial mass in a pendulum seismometer [see illustration on this page], the differential change in the level of mercury in a "tiltmeter" [top illustration on opposite page] and the changing distance between the piers of a "strainmeter" [bottom illustration on opposite page]. Optical interferometers and differential transformers also serve as transducers. The output from the transducer can be recorded not only as a wiggly line on a pen recorder or on photographic film but also in digital form, on magnetic tape that can be fed into a computer for subsequent analysis.

Modern seismographs are limited in sensitivity not by design considerations but by background noise. Microseisms, which originate in the transfer of energy from the wind or sea to the solid earth, can travel thousands of miles and register on an instrument. Industrial machinery is a source of noise; so is the diurnal heating and cooling of the outer few inches of the earth's crust. Even the periodic yielding of the solid earth under the influence of lunar tides can contaminate seismic signals. (One investigator's "noise" is of course another investigator's "signal"; earth tides are of particular interest to many geophysicists.)

The most sensitive pendulum seismographs can detect displacements of one millimicron, or a millionth of a millimeter. In an array of such instruments the detection capability increases—roughly as the square root of the number of elements. A huge array is currently being installed in Montana for the purpose of detecting small earthquakes and nuclear explosions. Its 500 seismographs, arranged in patterns covering

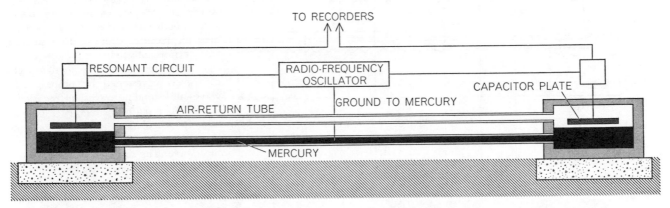

TILTMETER detects a tilt in the earth's surface by measuring the relative change in capacitance as the mercury level in two reservoirs changes. In this version designed by Hugo Benioff and William Gile of Cal Tech the reservoirs are three to 100 feet apart.

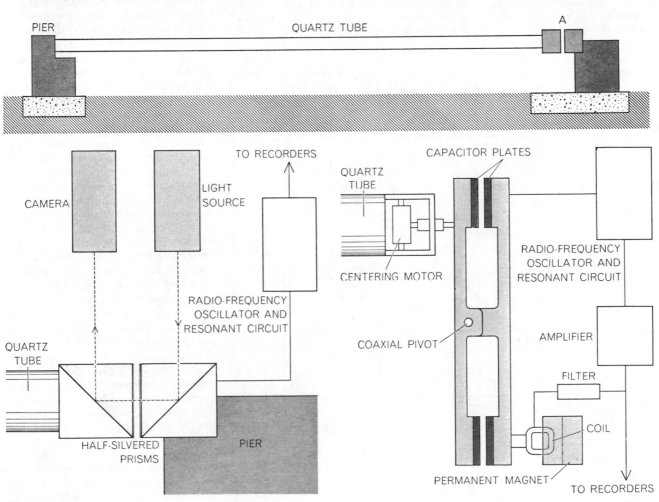

STRAINMETER designed by Benioff records strain, or a change in length over a given length, by measuring the changing distance between two piers. A quartz tube (*top*) extends from one pier almost to a second pier 10 to 100 feet away. A change in the small distance at *A* is measured by either of two transducers (*bottom*) designed with Leonard Blayney. In one (*left*) half-silvered prisms serve as capacitor plates and also constitute an interferometer that calibrates the instrument. In another transducer (*right*) the capacitor plates are pivoted and spurious signals are corrected for, as in the vertical pendulum seismograph, by a feedback system.

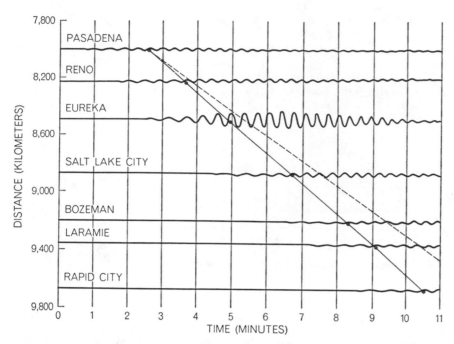

VELOCITY of Rayleigh waves is derived from seven seismograms recording the waves' arrival at instruments constituting a 2,000-kilometer array. The vertical scale shows distance from the epicenter. The solid line connects wave crests and indicates "phase velocity." The broken line connects wave groups of the same frequency and shows "group velocity."

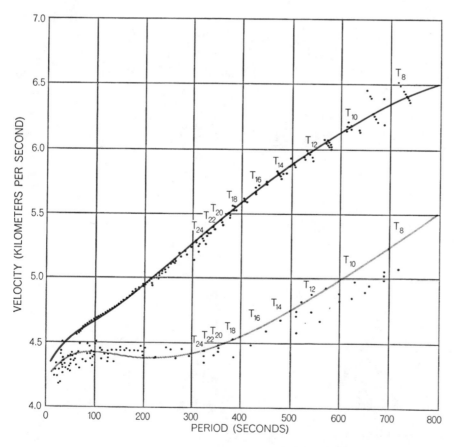

DISPERSION DATA, or phase and group velocities at various periods, were obtained experimentally from propagating Love waves for periods of less than about 300 seconds and from torsional free oscillations for longer periods. Data were plotted with theoretical curves of phase (*dark color*) and group (*light color*) velocity for a particular earth model.

an area 200 kilometers on a side, may sense ground motions smaller than a ten-millionth of a millimeter! A strain seismograph, which is inherently more sensitive to very long waves, measures deformation in terms of the change in length over a certain length. Strain-meters are limited by noise to sensing strains of 10^{-9}—say one millimeter over 1,000 kilometers—in the case of permanent displacements and of 10^{-10} in the case of transient waves.

I mentioned earlier the dual aspect of waves represented by propagating waves and standing waves, or free oscillations. As a matter of fact, each of the two free-oscillation modes is related to a specific kind of propagating surface wave. The torsional modes are the interference patterns of Love waves (named for A. E. H. Love, who discovered them); the spheroidal modes are the standing-wave equivalents of Rayleigh waves (named for Lord Rayleigh, who described them in 1900). Love waves are transverse shear waves: the ground vibrates horizontally at right angles to the direction of the wave. Rayleigh waves are like water waves: particles in their path vibrate in ellipses that lie in the vertical plane along which the wave is being propagated.

Free oscillations contain no information about the earth's interior that is not carried by the propagating surface waves. Whether one chooses to analyze a seismogram in terms of Love or Rayleigh waves or of free oscillations is therefore a matter of convenience. The length of the waves is the determining factor. Seismic waves radiate from a source and travel around the world repeatedly; a single wave arrives at a seismograph many times. In the case of waves with periods longer than about 1,000 seconds, with wavelengths of the order of the earth's radius, it is difficult to distinguish among these various arrivals, and such waves are best studied as free oscillations. Waves with periods shorter than 300 seconds or so are easily separated, however. Their lengths are shorter than about 1,500 kilometers and their speed is controlled by the properties of the earth's crust and upper mantle, properties that may differ below continents and ocean basins. By obtaining the propagation velocity of such waves in segments of their paths restricted to a single geological province, one can deduce crust and upper-mantle structure in detail. These shorter waves are therefore better studied as propagating waves than as global free oscilla-

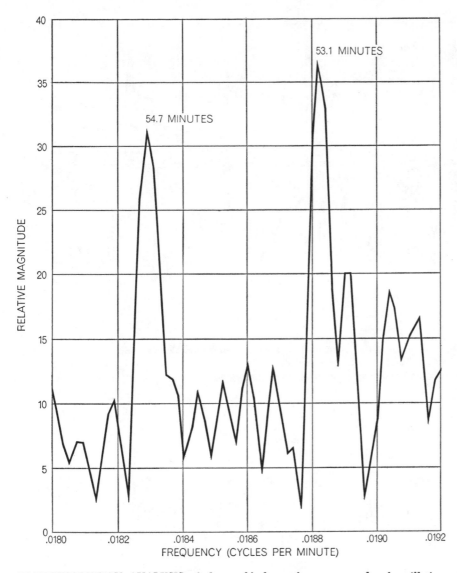

HIGH-RESOLUTION ANALYSIS of the earth's lowest-frequency mode of oscillation, the $_0S_2$ mode, revealed this double peak. Such multiplets are caused by the rotation of the earth, which affects the symmetry between waves traveling in opposite directions.

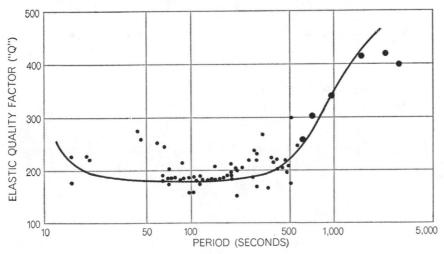

ELASTIC QUALITY, or the extent to which energy is retained in a system, is measured by the factor Q. Here Q is plotted as a function of wave period as determined from propagating-wave data (*small dots*) and free-oscillation data (*large dots*). The increasing Q at long periods implies that this factor increases with depth in the earth's mantle.

tions, which necessarily average out the geology of the entire world.

The propagation velocities of waves of various periods yield information on the structure of the earth because there is a correlation between the length of a wave and the depth it probes. Velocity data can be derived from the time at which waves arrive at a number of different stations [*see top illustration on page 24*], or arrive repeatedly at one station after circling the earth in both directions. The seismograms in the illustration record the arrival of Rayleigh waves from an earthquake in the Solomon Islands at seven stations in the U.S.; the seismograms are spaced according to the distance of the station from the epicenter and are aligned in time.

If a line is drawn connecting a point of constant phase in each seismogram, such as the crest of a certain wave, its slope gives the "phase velocity" of waves of a certain period. If, starting from the same point as the phase-velocity line, one connects instead the centers of groups of waves having about the same frequency, the slope of the line gives the "group velocity" for waves of the same period. Both kinds of velocity turn out to be greater for longer waves than for shorter waves; indeed, a glance at the seismograms shows that there are longer waves at the head of the wave train and shorter waves toward the rear. This dispersion, or sorting out, of the waves according to period—frequency modulation of the wave train, as it were—depends primarily on the variation of elastic velocity with depth in the earth: the long waves travel faster because they probe greater depths. Analysis of a seismogram for dispersion therefore tells one about the properties of various layers of the earth or of specific subcontinental or sub-oceanic regions, depending on the periods of the waves being studied.

Dispersion curves are built up from a large number of velocity observations [*see bottom illustration on page 25*]. One can also construct theoretical curves by working backward from assumptions about the earth's structure to the elastic velocities at various depths and thus to the speed with which waves of different lengths should propagate. There are theoretical curves that fit the observed data quite well. So far the fit is not unique because not enough data have yet been accumulated, but the available dispersion curves do at least limit the range of possible earth structures. Such

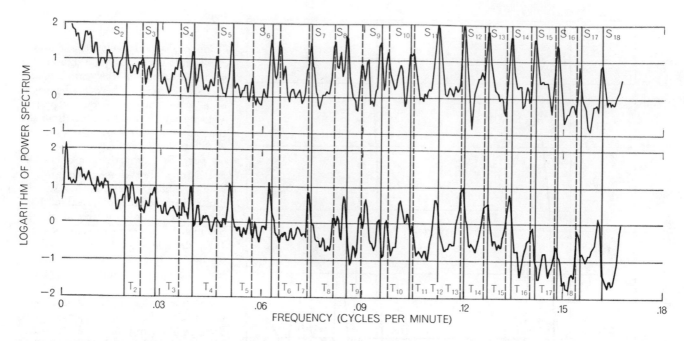

POWER SPECTRA of seismic waves from the Chilean earthquake of 1960 (*top*) and from the Alaskan earthquake of 1964 (*bottom*) are displayed together. The spectra were derived as explained in the text. Spectral peaks corresponding to spheroidal (*S*) and torsional (*T*) oscillations are labeled according to mode number. Peaks correspond in the two spectra; amplitudes differ.

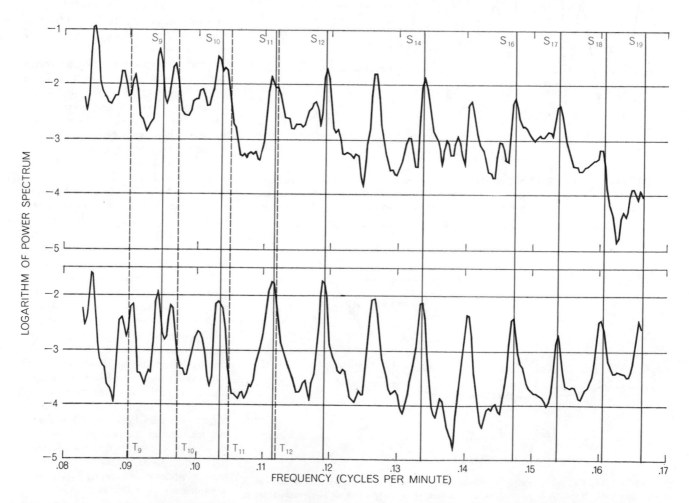

SPECTRA obtained from strainmeters in Peru (*top*) and in California (*bottom*) for the Chilean earthquake also show peaks that coincide in frequency as predicted by theory. "Noise" between peaks in the spectra, on the other hand, shows no cross-correlation.

constraints are important to the investigator; as constraints accumulate from different kinds of data they narrow the range of uncertainty, much as information from a number of senses converges to define an object.

Measurements of the earth's free oscillations have extended the depth to which one can examine the earth's structure. In the bottom illustration on page 24, for example, the data for waves with periods of less than about 300 seconds were obtained from measurements of Love waves. The data for longer waves, which reflect the properties of deep layers of the earth's mantle, came from analysis of the earth's torsional modes.

Although long surface waves have been studied intensively for a decade, the necessary tools for observing free oscillations and a sufficiently large seismic source to excite strong oscillations were not available at the same time until the great Chilean earthquake of May, 1960. Within a few weeks after that event teams from Cal Tech, Columbia University and the University of California at Los Angeles reported observing some 40 fundamental spheroidal modes, 25 torsional modes and several spheroidal overtones.

The experimental procedure for detecting free modes begins with a seismogram cast in the form of a time series—a record of the variation with time of the amount of strain or tilt or other motion of the earth. The output from the detecting instrument is quantified at discrete time intervals—say once every 10 seconds—and recorded on tape as a series of digits. The sequence of digits begins with the first signal from the earthquake and may last for a period of weeks or months, depending on how quickly the oscillations decay. (An analogue record is made simultaneously because it is easier to tell from a wiggly line than from a list of numbers whether or not the instruments are working.)

The problem of identifying free oscillations is simply one of decomposing the time series into its component modes. This is accomplished by the process of Fourier analysis. The numerical procedure, carried out in a computer, is to multiply the time series by a sine wave of a given frequency and to average the product over the entire series. The averaging procedure cancels components of all frequencies but the one being considered, and the average is proportional to the strength of that component in the series as a whole. The process is repeated at one frequency after another until the entire frequency band has been scanned. The result is a spectrum with peaks at the frequencies of the earth's normal modes of oscillation. The height of each peak is a measure of the power, or the degree of excitation, of the mode. One might liken Fourier analysis to the process of passing a complex electrical signal through a sharp filter. The filter can be adjusted to reject all frequencies but one, and its output is then proportional to the degree to which the harmonic component with that frequency is present in the original signal. The spectrum can be made cleaner by the process of cross-spectrum analysis. If two time series are derived from widely separated stations, and if the signal is correlated but the noise is not, the contamination of the spectrum by noise can be reduced significantly.

Stewart Smith of Cal Tech compared power spectra for the two great earthquakes of recent years, the one in Chile in 1960 and the one in Alaska last year. The earth's free-oscillation modes are plain to see from his data [see top illustration on page 26]. There is a sequence of spectral peaks, most of which are present in both spectra. As a result of the theoretical predictions by Pekeris and his colleagues it is possible to assign the appropriate S and T mode numbers to each peak. As theory requires, the modes reflect the properties of the earth, not the individual earthquake: the peaks fall at exactly the same periods whether they were excited by the Chilean or the Alaskan shock. The amount of energy in each mode, however, depends on the location and dimensions of the source, so that not all

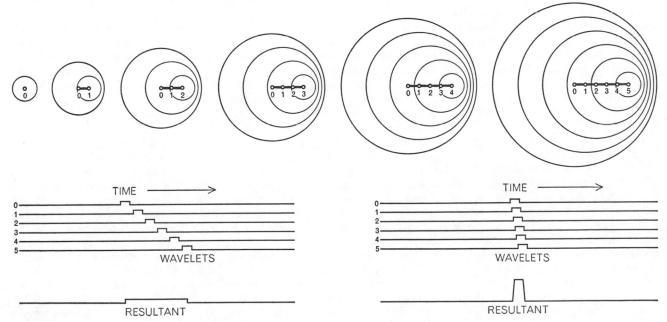

ASYMMETRIC WAVE PATTERN is radiated by an earthquake as it propagates along a fault. The top diagrams show how the waves excited at the first instant (0) and at five equal time intervals thereafter are crowded together in the direction in which the earth is cracking and are spread out in the opposite direction. Such wavelets will be recorded as resultants of greater amplitude and higher frequency at a seismograph placed in the direction of fault progression (bottom right) than at an instrument on the opposite side of the epicenter (bottom left). The ratio of spectra from two such instruments yields fault length and rupture velocity.

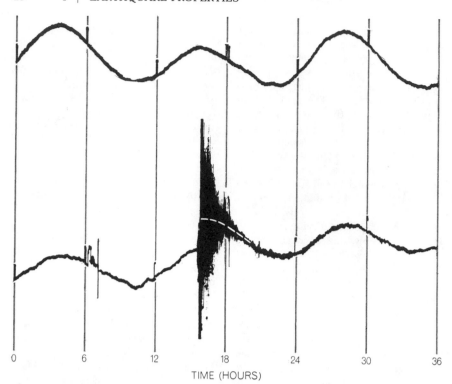

TIME (HOURS)

PERMANENT STRAIN CHANGE of 10^{-8} is indicated on a strain seismogram made in Hawaii for the Alaskan earthquake of 1964. The sinusoidal motion with a 12-hour period is due to earth tides. Waves from the earthquake are bunched solidly because of the recorder's slow speed. Broken white line is added to delineate the permanent strain.

the peaks are of the same magnitude in both spectra. By way of analogy, a violin string can be bowed heavily or lightly; it will sound the same fundamental frequency but will do so loudly or softly.

Just as the peaks coincide in spectra derived from two earthquakes recorded at one site, so they coincide in the case of spectra from two widely separated strainmeters measuring the same earthquake [*see bottom illustration on page 26*]. In this case the two curves were analyzed for coherence, a measure of correlation. The coherence turned out to be high in the vicinity of the peaks but low between the peaks; that is, the frequencies of the modes coincided but those of the noise in the two records did not. Another check with theory is the fact that T modes, which have no vertical component, do not show on spectra derived from vertical pendulum instruments, whereas both S and T modes appear on spectra from horizontal pendulums and strainmeters.

The theoretical prediction and experimental verification of the earth's free oscillations constitute one of the most elegant experiments in geophysics. As the spectra indicate, the mode with the lowest frequency, or the longest period, is the "football" mode, $_0S_2$. Its

mean observed period is 53.95 minutes. (In terms of what one might call the music of the spheres, that corresponds to E flat in the 20th octave below middle C!) This is amazingly close to the value of one hour Love predicted so long ago. As a matter of fact, the observed periods of all the S and T modes are extremely close to sets of theoretical values that had been worked out for four different earth models.

In the course of obtaining high-resolution spectra at Cal Tech we noted a curious phenomenon: many of the modes registered not as single peaks but as multiplets [*see top illustration on page 25*]. Pekeris in Israel and Freeman Gilbert and George Backus at the University of California at San Diego soon found that this was caused by the rotation of the earth, which makes the periods dependent—as they would not be in the case of a stationary earth—on the longitudinal index m. Rotation destroys the symmetry between waves traveling from east to west and those traveling from west to east. The excitation of the multiplets depends on the relative location of the source and the receiver: a seismograph at the South Pole would show no multiplets. Rotational splitting is of great theoretical interest but makes it harder to obtain

the precise values for modal frequencies that are needed to explore the earth's interior. "Geographic filtering" with data from many stations or even a special installation at the South Pole may be required to eliminate the effect.

The investigation of long waves and of free oscillations has given geophysicists a new tool for deducing the structure of the deep interior of the earth and for studying the mechanism of earthquakes. One new aspect of earth structure that is currently being studied is the degree to which various layers tend to dissipate energy. The factor that electrical engineers and investigators of materials call Q is a measure of dissipation in energy-storing systems; it is proportional to the energy stored in the system divided by the energy dissipated during a cycle. Electrical engineers obtain Q from the sharpness of the resonance curve in a circuit; materials scientists measure Q in free elastic vibrations. Solids with high Q values "ring." Typical values are 100,000 for quartz, 1,000 for steel, 100 to 1,000 for crystalline rocks. Q can also be characterized as the degree of perfection of elasticity, and one measure of it in the earth is the width of spectral peaks in the free oscillations. The decay of Love and Rayleigh waves after repeated circling of the earth is an alternative measure.

Using both methods, Don L. Anderson and Charles Archambeau of Cal Tech showed that the increase in Q with period corresponds to an increase in Q with depth in the mantle [*see bottom illustration on page 25*]. The longer waves sample the deeper portions of the mantle, where the rocks are surprisingly less dissipative, or more perfectly elastic. The Q of rocks increases with pressure, decreases with temperature and depends on the rocks' physical state and composition. Since both pressure and temperature increase with depth, geophysicists are trying to decide if the low Q in the upper mantle indicates that temperature is more effective than pressure at these depths; in that case the reverse would have to be true at greater depths, where Q is of the order of 1,000. Alternatively, the mantle might be partially melted in the low-Q zone. Does a low Q correlate with low strength? The weak upper mantle this implies would account for the buoyancy by which mountains are supported. It would explain the large-scale horizontal movements of the earth's crust that seem to be indicated by studies of the directions in which ancient rocks were magnetized. The low-Q area, moreover, seems to coincide with a zone in the upper mantle of low shear velocity, indicated by recent dispersion data. The reason for the unexpected decrease in velocity could be that the rate at which velocity decreases with temperature is simply greater, at this point on the scales, than the rate at which it increases with pressure. On the other hand, it could imply that temperatures in the upper mantle are higher than had been thought and that a fraction of the mantle rock is melted. Geologists have speculated that there is a zone of molten basaltic rock at this level.

Mark Landisman and Anderson have suggested another application of free-oscillation data: the derivation of the density distribution of the earth. The present method for deducing densities requires arbitrary assumptions as to chemical composition and other factors. It appears that when the effect of velocity has been allowed for, the free oscillation periods are particularly sensitive to density, and studies are now under way in which the new techniques are combined with data on the overall velocity distribution of the earth and the earth's mass and inertia.

Seismologists would like to find out just what goes on at the source of an earthquake—how long the fault is, for example, and how quickly it propagates. Benioff and Ari Ben-Menahem of Cal Tech suggested recently that an earthquake could be considered as a traveling, radiating disturbance and that the resulting asymmetry of its radiation pattern could be a clue to the length of the fault and the velocity of rupture. Detectors on opposite sides of a fault receive different signals as the crack in the earth propagates [*see illustration on page 27*]. On one side the waves are crowded together; the pulse is larger and of higher frequency. On the other side the amplitude is less and the wave is longer. Ben-Menahem suggested dividing the free-oscillation spectrum of the waves on one side by the spectrum of the waves on the other side; the ratio should depend on the length of the fault and the velocity of rupture. Ben-Menahem went on to cast the idea in mathematical form, defining a "directivity function" by the application of which it has been possible to determine the length and rupture velocities of a number of earthquake faults. In the case of the Chilean earthquake, for example,

the fault was about 750 kilometers long and the crack propagated at between three and a half and four and a half kilometers per second.

The longest possible elastic wave in the earth is the $_0S_2$ mode, with a period of about 54 minutes. (In anticipation of seismic measurements on the moon in the not too distant future, incidentally, Bruce A. Bolt of the University of California at Berkeley has calculated a 15-minute period for the same mode for one model of the moon.) There are actually "waves" of infinitely long period: permanent deformations of the earth. I found that the Alaskan earthquake, for example, caused a permanent strain change of 10^{-8} in Hawaii. The strain is small, but since it was detected several thousand kilometers from the source [*see illustration on opposite page*] it must represent a permanent displacement of the order of a centimeter. Such effects suggest a new field of study that has been dubbed "zero-frequency seismology." It holds real promise as a means of learning about the actual mechanism of an earthquake and defining the relation of earthquakes to mountain-building and continental drift.

QUARTZ ROD of a strain seismograph installed at the Cal Tech facility on Oahu, Hawaii, is shown in this photograph. The rod is about 100 feet long and has measured strains as small as 10^{-8}. The seismogram on the opposite page was recorded by this instrument.

3 Earthquake Prediction

by Frank Press
May 1975

Recent technical advances have brought this long-sought goal within reach. With adequate funding several countries, including the U.S., could achieve reliable long-term and short-term forecasts in a decade

The forecasting of catastrophe is an ancient and respected occupation. It is only in recent years, however, that earthquake prediction has parted company with soothsaying and astrology to become a scientifically rigorous pursuit. At present hundreds of geophysicists and geologists, mainly in the U.S., the U.S.S.R., Japan and China, are engaged in research with earthquake prediction as the direct goal. Most of these investigators believe that the goal is attainable. Some are more pessimistic. A few actually think that the side effects of prediction might be worse than the benefits and that the goal should be abandoned. Research on earthquake prediction therefore exemplifies many of the problems that face modern society: technology assessment, the design and organization of a massive mission-oriented project, the competition for funds and the political niceties of an undertaking involving admittance to previously inaccessible regions of another country.

I share the view of most of my colleagues that earthquake prediction is a highly desirable goal. Because of the large increase of population density in the earthquake-prone sections of the U.S., the potential loss from an earthquake as strong as the San Francisco shocks of 1906 could be as high as tens of thousands dead and hundreds of thousands injured, with property damage measured in the billions of dollars. A catastrophe on this scale would be unprecedented in the history of the country, yet it is an event that most seismologists expect to occur sooner or later. The seismic-risk map of the U.S. shows the most probable locations of strong earthquakes [*see illustration at left*]. The map is based primarily on earthquake history; it does not take into account the frequency of occurrence. Hence Boston is shown to be as risky as Los Angeles (mainly because of a single great quake that occurred in the Boston area in 1755), even though tremors are 10 times less frequent on the East Coast than on the West Coast. It is a sobering thought that a third of the nation's population live in the two regions of highest risk.

Preliminary results of current investigations indicate that predictions of strong earthquakes could be made many years in advance. It also appears likely that a method for making short-term

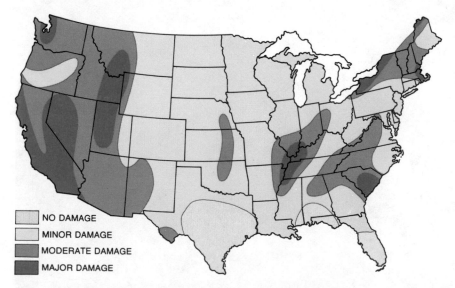

NO DAMAGE
MINOR DAMAGE
MODERATE DAMAGE
MAJOR DAMAGE

RISK OF DAMAGE FROM EARTHQUAKES is assessed in a broad, climatological sense in this map of the U.S., based on information compiled by the Coast and Geodetic Survey. The map is based primarily on historical records of destructive earthquakes and does not take into account the fact that earthquake tremors are much more frequent in the Western states than elsewhere. A third of the nation's population live in two darkest-colored regions.

predictions, as short as weeks or even days, will be developed. With this dual capability it should become possible to devise a remedial strategy that could greatly reduce casualties and lower property damage. For example, the long-range prediction of a specific event could spur the strengthening of existing structures in the threatened area and motivate authorities there to enforce current building and land-use regulations and to revise such codes for new construction. A public-education campaign on safety procedures could also be instituted.

Short-term prediction could mobilize disaster-relief operations and set in motion procedures for the evacuation of weak structures or particularly flammable or otherwise hazardous areas. The shutdown of special facilities, such as nuclear power plants and gas pipelines, and the evacuation of low-lying coastal areas subject to tsunamis, or "tidal waves," could also follow a short-term forecast.

The problem of how one communicates an earthquake prediction to the public and the consequences that flow from such warnings (and from possible false alarms) are now being examined. Research into the social aspects of earthquake prediction will presumably advance along with progress toward a physical solution of the problem. For these reasons most experts consider the ability to predict earthquakes to be justifiable on both humanitarian and economic grounds.

With the advent of the theory of plate tectonics the distribution of earthquake belts around the world became understandable. According to this view, the earth's lithosphere, or outer shell, is divided into perhaps a dozen rigid plates that move with respect to one another. Most of the large-scale active processes of geology—vulcanism, mountain-building, the formation of oceanic trenches, earthquakes—are concentrated at or near plate boundaries [*see illustration on next two pages*]. It is easy to see why stresses build up along plate boundaries, where the relative motion of the plates is resisted by frictional forces. When the stress increases to the point where it exceeds the strength of the rocks of the lithosphere or overcomes the frictional forces at the boundary of a plate, fracturing occurs and an earthquake results. The plate-tectonic model combined with earthquake statistics already makes it possible to predict earthquakes in the climatological sense of identifying particularly dangerous areas

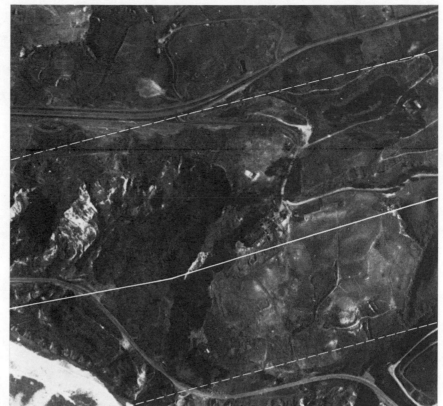

HOUSING TRACTS constructed within the San Andreas Fault zone near San Francisco appear in the aerial photograph at bottom, made by Robert E. Wallace of the Geological Survey in 1966; the photograph at top shows the same scene approximately 10 years earlier. Solid white line in each view traces the approximate position of fault along which the ground ruptured and slipped some two meters during the great earthquake of 1906. Broken white lines give approximate boundaries of main fault zone. Pacific Ocean is at lower left.

and estimating the relative degree of danger. What is needed, however, is prediction more akin to weather forecasting: Where and when is the next earthquake likely to take place?

A combination of laboratory and field experiments over the past five years has led to a breakthrough in thinking about the problem of earthquake prediction. When a rock is squeezed, it deforms and eventually breaks. Just before it breaks it swells, owing to the opening and extension of tiny cracks. This inelastic increase in volume, a phenomenon long known to laboratory experimenters as dilatancy, begins when the stress reaches about half the breaking strength of the rock. In the mid-1960's William F. Brace and his colleagues at the Massachusetts Institute of Technology showed that measurable physical changes accompany dilatancy in laboratory experiments;

such effects include changes in the electrical resistivity of the rock and in the velocity at which elastic waves travel through the rock. Brace suggested that dilatancy and its effects might be detectable in the earth's crust and provide a basis for earthquake prediction; his suggestion generated much excitement at the time because it opened up the possibility that premonitory physical changes could be observed in advance of earthquakes.

In the late 1960's two Russian investigators, A. N. Semenov and I. L. Nersesov, startled the seismological world with a report that unusual variations in the velocity of seismic waves appeared just before earthquakes in the Garm region of Tadzhikistan. Subsequently the Russians announced that in earthquake-epicenter regions in Garm, Tashkent and Kamchatka they had detected changes

both in electrical resistivity and in the content of the radioactive gas radon in the water of deep wells.

These reports triggered a flurry of activity in the U.S. American seismologists went to the U.S.S.R. to see the data firsthand. They also began arranging their own experiments in order to observe the precursory phenomena. Technical papers on these phenomena authored by Russian, American and Japanese workers began to be presented in increasing numbers at scientific meetings and in journals. Last year a group of American geologists and geophysicists visited China and found a large-scale earthquake-prediction program under way, with important results that had not yet been reported at international meetings or in publications.

It is fortunate that a number of earthquake precursors have been found, each

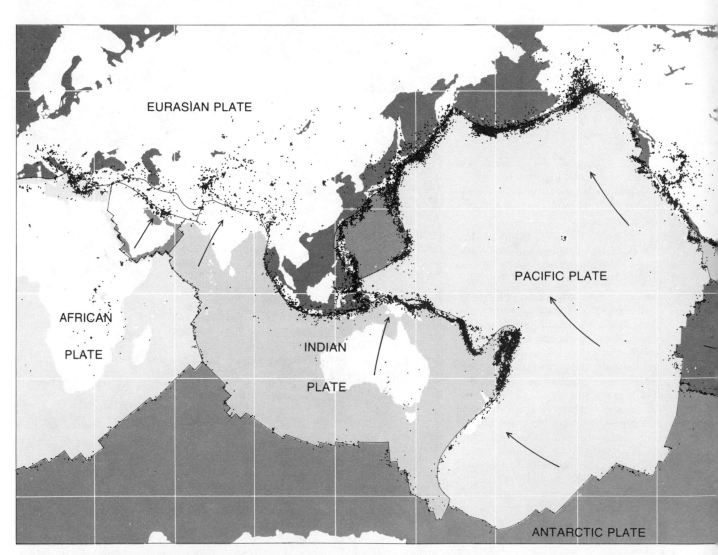

EPICENTERS OF 30,000 EARTHQUAKES recorded between 1961 and 1967 are indicated on this world map plotted by M. Barazangi and H. J. Dorman of Columbia University on the basis of information supplied by the Coast and Geodetic Survey. Also shown are the dozen or so moving plates that, according to the modern the- ory of plate tectonics, comprise the earth's rigid outer shell. Most earthquakes take place at or near plate boundaries, where the relative motion of the plates is resisted by frictional sticking until the stress builds up to the point where the rock fractures, causing an earthquake. Intraplate earthquakes, such as those that appear in the

based on a different physical measurement. Confidence in a prediction is enhanced when it is based on several independent lines of evidence, each with its distinctive "noise" history and its distinctive anomaly signaling an earthquake. How are the precursory anomalies observed?

An array of seismographs can be used to sense precursory changes in the velocity of compressional waves and shear waves in the focal region of an earthquake [see illustration on next page]. The seismic waves originate in smaller earthquakes within the focal region, in larger earthquakes outside the focal region or in artificial sources such as explosions or mechanical devices. Such anomalous changes have been observed in several parts of the U.S., the U.S.S.R. and China.

Seismically active regions have many

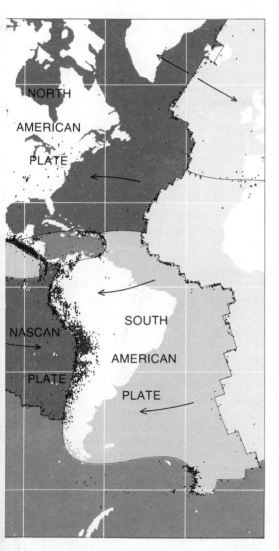

eastern U.S., are rare but can be destructive. China, squeezed by large plates on the south and the east, has a high level of seismic activity, which may be attributable to the existence of "miniplates" in central Asia.

more small earthquakes than large ones. This "background" of small tremors varies in time. Periods of calm before a strong shock are frequently observed; the background activity appears to go through a minimum and then to increase just before the main shock. The pattern of radiation of seismic waves reflects the stress field in the crust. In central Asia Russian investigators have found that the stress pattern shown by the small tremors is random during the calm period but becomes highly organized beginning three or four months before the main shock. The compressional stresses become aligned in the same direction as that of the forthcoming main shock.

Another approach is to measure anomalous changes in the volume of crustal rock in the focal region. The changes can be observed by tiltmeters, by devices for monitoring changes in sea level (corrected for oceanographic and meteorological effects) and by repeated surveying. In parts of Japan and China historical records of precursory changes in the level of lakes, rivers or the sea, sometimes dating back hundreds of years, may be related to the same phenomena [see illustration on page 35].

Precursory changes in water level, water turbidity and temperature in deep wells can be observed visually or with instruments. Observing the radon content of well water, a technique used extensively in the U.S.S.R. and China, also seems to be a sensitive indicator of forthcoming seismic activity [see top illustration on page 37].

If an electric current is fed into the earth's crust between two points several kilometers apart, voltage changes between two other points will show up if the resistivity of the intervening crustal rocks changes. Such precursory fluctuations have been reported in the U.S., the U.S.S.R. and China [see bottom illustration on page 37].

Magnetometers on the earth's surface can detect changes in magnetic field with a strength of about a hundred-thousandth of the earth's natural field. By subtracting the changes sensed by "standard" instruments removed from the epicentral region, noise introduced by fluctuations in the stream of electrically charged particles from the sun (the "solar wind") can be reduced and anomalous changes in the focal region can be detected. Precursory magnetic signals have also been observed in the U.S., the U.S.S.R. and China.

Although one can conceive of an earthquake-prediction strategy based purely on empirical observations such as

these, it is highly desirable to have a physical model that explains the observations. A model not only enhances confidence in the basic notion of predictability but also makes for more efficient research procedures.

Two principal models have been proposed, both growing out of laboratory experiments. The dilatancy-diffusion theory, proposed by Amos M. Nur of Stanford University in 1972 and extended by Christopher H. Scholz, Lynn R. Sykes and Y. P. Aggarwal of Columbia University in 1973, is supported by most American specialists. Another model, which might be called the dilatancy-instability theory, was proposed in 1971 by workers at the Institute of Physics of the Earth in Moscow. It also has a few American and Japanese adherents. The models have a common feature: the growth of cracks as stress builds up in the crust just before an earthquake [see illustration on page 39].

Both models begin with a stage in which elastic strain builds up in the earth's crust. In the next stage small cracks open in the strained portion of the crust and dilatancy becomes a dominant factor. In the Russian view the development of cracks "avalanches" in this stage. In both models it is the second stage that marks the real beginning of precursory phenomena, since the open cracks change the physical properties of the rock. Seismic velocity (the ratio of compressional-wave velocity to shear-wave velocity) drops. Electrical resistivity increases if the rock is dry and decreases if it is wet. Water flow through the rock increases (and therefore more radon enters the water from the rock). Volume in the dilatant zone increases. In the American model the number of small tremors decreases in this stage because the cracks become undersaturated in water as they increase in number; as a result sliding friction increases and inhibits faulting.

The two models differ markedly in the third stage. In the American model water diffuses into the undersaturated dilatant region. The main effect of this inflow is to increase the seismic velocity and to raise the pore pressure in the cracks, weakening the rock to the point where small earthquakes increase in number and the main shock follows. In the Russian model water plays no role in the third stage. Instead the avalanche-like growth of cracks leads to instability and rapid deformation in the vicinity of the main fault. The stress load drops partially in the region surrounding the zone of unstable deformation, cracks partially close and the rock recovers some of

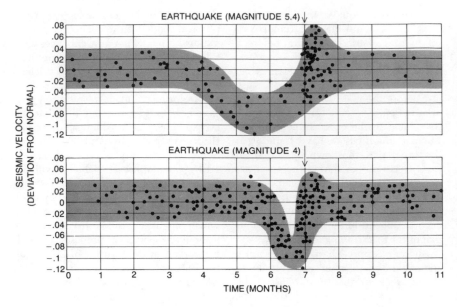

PREMONITORY CHANGES in seismic velocity (the ratio of compressional-wave velocity to shear-wave velocity) were observed in the late 1960's just before two fairly large earthquakes in the Garm region of Tadzhikistan by A. N. Semenov and I. L. Nersesov of the Institute of Physics of the Earth in Moscow. These composite diagrams, drawn from their work, are based on a number of smaller earthquakes in the region. Each point represents a deviation of the seismic velocity from the normal regional value and is derived by measuring the travel times of compressional waves and shear waves from each small earthquake to a local network of seismograph stations. The colored bands indicate the statistical scatter of the observations. The duration of the calm period preceding the main earthquake appears to increase with the magnitude of the forthcoming event. (The two earthquakes shown measured 5.4 and 4 on the Richter scale.) Seismic-velocity anomalies of this type have been observed about 18 times in U.S.S.R., 10 times in U.S. and several times in China.

its original characteristics. This sequence of events accounts for the increase in seismic velocity, the decrease in volume and the other changes typically observed in the third stage. The developing instability finally gives way to faulting, and the main shock ensues. In both models stress is released by the earthquake, and the crustal rock recovers most of its original properties.

An empirical formula, derived by James H. Whitcomb, J. D. Garmany and Don L. Anderson of the California Institute of Technology, connects the duration of the precursory anomaly with the magnitude of the predicted earthquake. For example, an event with a magnitude of 5 on the Richter scale has an anomaly lasting for about four months, whereas a major earthquake, with a magnitude of 7, say, would be preceded by an anomaly beginning some 14 years before the event. The formula is still rough, particularly in the high-magnitude range, but it appears that the large earthquakes will provide warning times on the order of 10 years. The discovery that the size of an earthquake, as well as its location and timing, is predictable should hold important implications for the design of an earthquake-mitigation strategy. Fortunately the larger the magnitude of the

forthcoming quake, the longer the lead time available for making plans to combat its effects.

What is most needed now to bring earthquake-prediction technology to the point of implementation is a larger number of examples of successful prediction. So far only about 10 earthquakes have been predicted before the fact. Perhaps three times as many have been "predicted" after the fact by going back to the data and finding precursory signals. It is difficult to know how many formal predictions, based on the methods described above, have failed. The number is probably less than 10, which is not bad for the rudimentary research networks now in operation. That is still too small a sample to eliminate unreliable methods and to design a comprehensive, operating prediction system. Although the major earthquake belts extend for tens of thousands of kilometers, only a small fraction of that distance is instrumented adequately to test prediction methods. With the pooling of the data being gathered in various countries, however, the number of case histories should grow rapidly in the next few years, and statistically valid tests of prediction methods should be forthcoming.

Seismologists of different nationalities, like workers in any other field of science, need to combine their results in order to advance toward a common goal.

The leading agency for earthquake-prediction research in the U.S. is the Geological Survey, which runs a strong program centered in California and supports a research program in several universities. In central California, the region where the San Andreas Fault is most active, the Geological Survey has installed a network of stations equipped with seismometers and tiltmeters. Magnetic and electrical observations are also conducted but to a much lesser degree. In southern California a large number of instruments are being installed in a joint effort involving the Geological Survey and Cal Tech. Data from these arrays are mostly telemetered into Menlo Park and Pasadena on telephone and microwave circuits. This growing ability to pinpoint earthquake locations and monitor precursory velocity changes, tilts, magnetic fluctuations and changes in electrical resistivity is beginning to pay out. Recently workers associated with the Geological Survey found that 10 California earthquakes were preceded by tilt changes in the vicinity of the epicenter [*see illustrations on page 36*]. Precursory changes in seismic velocity have been reported for about 10 earthquakes in California and New York.

Perhaps the most significant new data were gathered on November 28, 1974, when a magnitude-5 earthquake struck about 10 miles north of Hollister in central California. The tremor was preceded by distinct tilt changes and magnetic fluctuations convincingly above the noise level, and with indications of seismic-velocity changes. John H. Healy of the Geological Survey chided his colleagues the night before the quake for not publicly announcing the forthcoming event.

In spite of these interesting results the U.S. program is still not sufficiently supported to make prediction a reality within the next decade. It is simply a matter of too few methods being tested in too few places. With the present level of support many potentially important methods cannot be tested, such as arrays of wells monitoring water level and radon content, networks of resistivity sensors, sea-level gauges, advanced surveying techniques and so forth. Even now more data are being accumulated than can be digested, a situation that could easily be rectified if a large computer were provided to scan and automatically analyze the incoming stream of data. Universities and industries with much research talent are insufficiently involved because of the lack of funds. Few studies are being conducted outside California. An additional $30 million per year could make prediction within a decade a realistic goal. The cost-effectiveness of such an investment is obvious when one remembers that the relatively modest San Fernando tremor (magnitude 6.6) that struck just north of Los Angeles in 1971 resulted in damage of more than $500 million.

The earthquake-prediction program of the U.S.S.R. is centered in the Institute of Physics of the Earth in Moscow. A program involving laboratory and field measurements, comparable in level to our own, is being carried out. The Russian field experiments form the longest series so far, having been started nearly 20 years ago. The impressive discovery of anomalous precursors stems from these efforts. The strategy of the Russian investigators is somewhat different from our own in that several experimental sites are being monitored in central Asia and Kamchatka with a lower density of instruments compared with our heavier emphasis on specific areas in California. Moreover, the Russians are exploring more methods than we are. Nevertheless, it appears that in the absence of a major new initiative, operat-

ANOMALOUS UPLIFT of the earth's crust in the vicinity of Niigata in Japan was observed for about 10 years before the disastrous 7.5-magnitude earthquake there in 1964, according to the Japanese investigator T. Dambara. The uplift was detected by plotting changes in the height of bench marks measured in repeated land surveys. The graphs at right correspond to the lettered bench-mark sites (*black dots*) shown on the detailed map. Evidence of the crustal uplift was also obtained from records showing a precursory drop in mean sea level observed by a tide-gauge station at Nezugasaki.

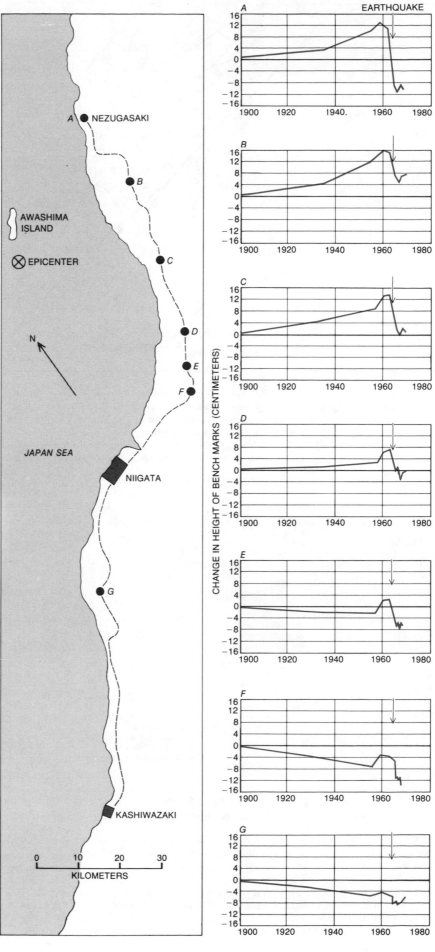

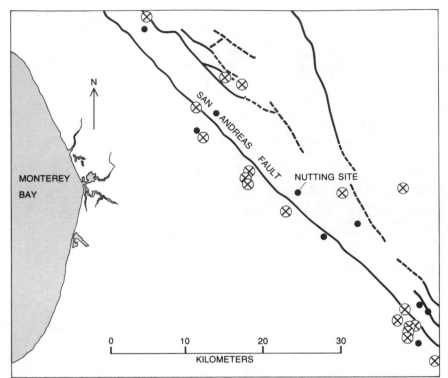

TILTING OF EARTH'S CRUST just before earthquakes has been observed by investigators associated with the Geological Survey using an array of sensitive tiltmeters (*black dots*) installed along 85 kilometers of the San Andreas Fault east of Monterey Bay. Circled crosses denote epicenters of all earthquakes with a magnitude greater than 2.5 recorded in the region between July, 1973, and March, 1974. Data summarized in illustration below were obtained at Nutting tiltmeter site, seven kilometers southwest of the town of Hollister.

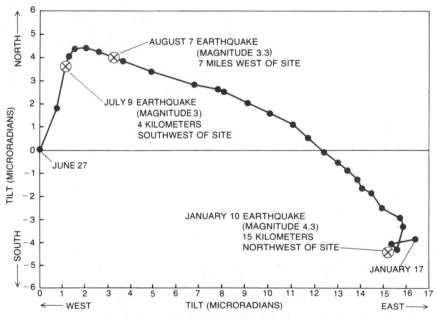

SEVEN-MONTH RECORD of crustal tilting was made during parts of 1973 and 1974 with the aid of a tiltmeter located in a shallow hole at the Nutting site. The colored dots represent the weekly mean tilt direction and magnitude. Several major local earthquakes are indicated; each is preceded by clear precursory change in tilt direction. M. J. S. Johnston and C. F. Mortensen of the Geological Survey report that precursory anomalies of this type have been detected on at least 10 occasions; the largest such event was on November 28, 1974, when a magnitude-5 earthquake struck about 10 miles north of Hollister. So far no comparable tilt change has been recorded that has not been followed by an earthquake.

ing prediction systems covering large areas will not be forthcoming in the U.S.S.R. either in this decade. As part of the environmental treaty between the U.S. and the U.S.S.R. there has been a rewarding exchange of ideas and personnel in the fields of earthquake prediction and seismic engineering. A formal bilateral working group has been established. In this way both American and Russian workers are kept informed of the latest unpublished developments; joint experiments are under way, and there is healthy criticism of each side's efforts by the other. This kind of close cooperation would have been unthinkable a few years ago.

Although Japanese earth scientists have been devoted to the notion of earthquake prediction since the turn of the century, a formal research program dedicated to this goal did not get under way until 1965. For years reports of anomalous sea-level changes and tilts prior to earthquakes have emanated from Japan, but the data were sparse and of uneven quality, and the world community of geophysicists was unimpressed. It now seems that some of these reports must have described true precursory phenomena. In any case the Japanese workers include some of the world's best geophysicists. It is therefore a tragedy that a strike has crippled the Earthquake Research Institute in Tokyo for several years.

The Japanese are currently emphasizing surveys every five years extending more than 20,000 kilometers. So far 17 observatories have been equipped with strain detectors and tiltmeters. Observations of the level of seismicity, of changes in the velocity of seismic waves and of magnetic and electrical phenomena are also under way. Cooperation between the U.S. and Japan in this field is quite close.

This past October I had the good fortune to participate in a month-long trip to China as a member of a group of 10 American earthquake specialists. This tour of Chinese research facilities followed a visit by 10 Chinese earthquake experts to the U.S. earlier last year. Since scholarly publication in China was suspended during the "cultural revolution," almost everything we saw in China was new to us. Following the destructive Hsing-t'ai earthquake of 1966 the Chinese embarked on a major effort in the field of earthquake prediction. Chairman Mao and Premier Chou En-lai issued statements charging Chinese scientists with achieving this goal. At present some 10,000 scientists, engi-

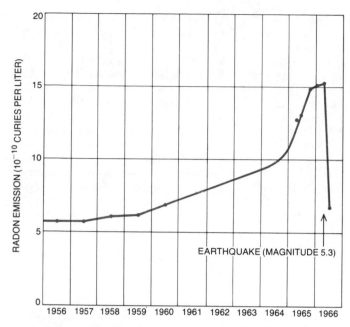

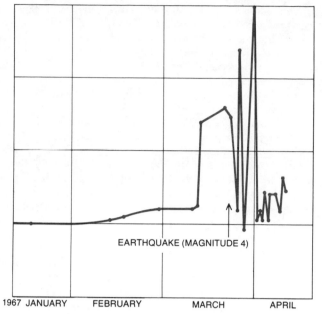

AMOUNT OF RADIOACTIVE GAS RADON dissolved in the water of deep wells has been found by Russian researchers to increase significantly in the period preceding an earthquake. The two examples shown here were recorded before two major earthquakes in the vicinity of Tashkent. The 1966 event (*left*) had a magnitude of 5.3; the 1967 aftershock (*right*) had a magnitude of 4. This promising observational technique is used extensively in both the U.S.S.R. and China, but it has not yet been tried in U.S.

neers, technicians and other workers are engaged in the program—more than 10 times the number of such workers in the U.S.

A unique feature of the Chinese approach is the use of an even larger number of amateurs, mostly students and peasants, who build their own equipment, operate professional instruments in remote areas and educate the local people about earthquakes. So far 17 fully equipped seismograph stations and

250 auxiliary stations have been installed. Data pertinent to earthquake prediction are being obtained at a total of 5,000 points. Every method described in this article is being tested in China. The Chinese say that they have made successful predictions, involving the evacuation of people from their homes and a consequent saving of lives. They also admit to false alarms and failures, chalking these up to the fact that their program is new and they are still in a

learning phase. The motivation for success is strong. The high population density, the nature of rural construction and the high degree of seismicity make China particularly vulnerable to earthquakes.

Although it is difficult to gauge the quality of the Chinese program from a brief visit, there is no question that the potential is great. In a few years the Chinese will probably be gathering more data than anyone else, owing to the size of their program and the more frequent

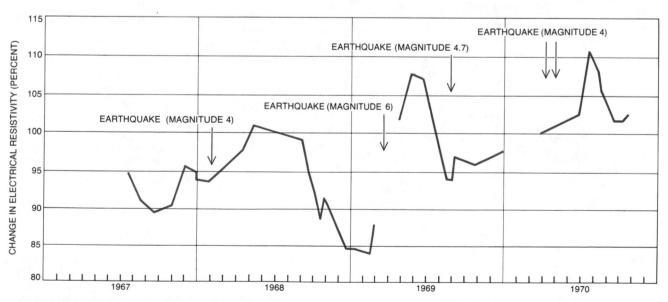

CHANGES IN ELECTRICAL RESISTIVITY of the earth's crust prior to earthquakes have been reported in the U.S.S.R., China and the U.S. The data for this graph were obtained by G. A. Sobolev and O. M. Barsukov for a series of earthquakes monitored in the U.S.S.R. between 1967 and 1970. Measurements of this type are made by feeding an electric current into the ground and observing voltage changes a few kilometers away. In general it has been found that earthquakes are preceded by a decrease in crustal resistivity.

incidence of earthquakes in their country. It may well be that the first statistical validation of prediction methods will come from China. It would be a pity if political considerations were to inhibit close international cooperation on this score, because joint projects with China could pay out in a more rapid achievement of a mutually desirable goal.

Although the prediction of earthquakes has been emphasized in this account, a comprehensive program to reduce vulnerability to destructive earthquakes includes progress in other areas: earthquake engineering, risk analysis, land-use regulation, building codes and disaster preparedness. Unlike earthquake prediction, about which there is a good deal of optimism but (so far at least) no guarantee of success, research and development in these other areas is bound to result in reduced casualties and lowered economic losses. Earthquake engineering deals with the efficient and economic design of structures that may have to withstand the shaking of earthquakes. The alteration of existing structures to improve their performance is included. Not only are residences, commercial buildings, schools, hospitals, dams, bridges and power plants examined individually but also

the interaction of all these elements in the system we call a community is considered. This developing technology can serve its purpose only if it is transferred from the investigators to the professional practitioners and to the regulatory bodies that draft building codes.

The damage caused by recent earthquakes in Japan and Alaska dramatized the fact that structures that could withstand the shaking were nevertheless toppled by foundation failure. Severe ground-shaking can cause soils to settle or liquefy and thereby lose their ability to support structures. Research on this poorly understood phenomenon is an important aspect of earthquake engineering. When it is better understood, it might be possible to take countermeasures or to institute land-use regulations that would limit construction on vulnerable soils as well as along active faults, in potential landslide areas or in coastal zones subject to tsunamis.

Some regions suffer major earthquakes frequently, others suffer them infrequently. In some places the potential for severe ground-shaking is higher because seismic waves propagate with less attenuation or because the soil resonates and amplifies the ground motion. In one city the problem following a quake is

fire; in another it is flooding. Construction practices differ from region to region. Some of these factors are known explicitly; some can only be described in probabilistic terms. All of them, and other factors as well, must be combined into an overall assessment of risk on which decisions must be based. Risk assessment is a new and important part of earthquake research. Also helping to provide a rational basis for decision making about land use and construction in earthquake regions is economic analysis of such questions as to what degree, in an area with a given probability of strong earthquakes, the added costs of safer construction are offset by the potential saving of life, property and productivity.

The possibility of controlling or modifying earthquakes arose a few years ago as the result of a chance discovery. The injection of wastewater into a deep well near Denver was found to have triggered small earthquakes. Since that time both laboratory and field experiments have shown that the injection of a fluid in a fault zone reduces frictional resistance by decreasing the effective normal stress across the fault. In a sense fluid injection serves to weaken the fault, whereas fluid withdrawal can strengthen it. If a preexisting stress is present, an earthquake could result if a fault were unlocked by fluid injection. In a remarkable field test of these ideas workers from the Geological Survey injected and withdrew fluids in a water-injection well of the Rangeley oil fields in Colorado and found that in this way they could switch seismicity on and off.

The extension of these results to the control of a major active fault such as the San Andreas Fault is unlikely in the near future. Some future generation, however, may be able to modify earthquakes by the injection of fluid and the controlled, gradual release of crustal strain. Science often advances more rapidly than is expected, of course, and in any case research on the possibility of modifying earthquakes should be encouraged for the sake of the next generation, if not of our own.

Although the number of case histories is still too small to make a positive statement about the feasibility of earthquake prediction, most seismologists would agree that prediction is an achievable goal in the not too distant future. Unfortunately the level of present effort in the U.S. is below that required to move rapidly to an operating prediction system. If a major earthquake were to

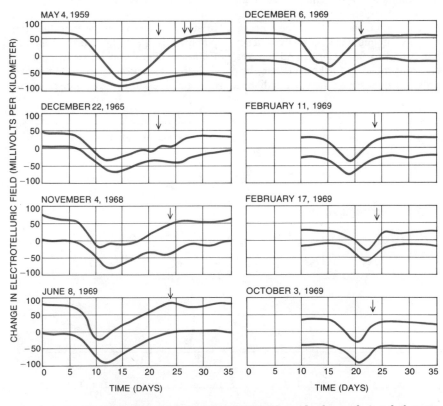

REDUCTION IN NATURAL ELECTRIC CURRENTS inside the earth just before an earthquake has also been observed by the Russian investigators. Data were obtained by recording voltage changes between points a few kilometers apart. Arrows denote earthquakes.

strike the U.S., the following day would almost certainly see abundant resources made available for a large-scale earthquake-mitigation program. (The earthquake-prediction programs of China and the U.S.S.R. were launched after severe earthquakes in each of those countries.) How does one sell preventive medicine for a future affliction to government agencies beleaguered with current illnesses?

It is proper, I believe, for scientists to assume an advocate role when they perceive an inadequate government response to some new opportunity or to some future danger. Earth scientists have a case to make. They can point to housing tracts placed in fault zones or on unstable hillside slopes. They can cite a newly built hospital that collapsed when shaken by the moderate San Fernando earthquake. The same tremor caused a dam to be stressed to near the failure point. A slightly larger shock would have resulted in casualties in the tens of thousands in the floodplain below the dam. Scientists can question the policy of a government that spends billions in construction but is unable to support research that would safeguard its own investment. They can question the wisdom of budgeting less than a tenth of a percent of the total construction investment for research on possible hazards. They can show how a research dollar invested today can yield an enormous return in lives saved and property preserved tomorrow. At a time when basic research budgets have not kept pace with the growth of the economy as a whole, earth scientists can point up the practical value to society of their new comprehension of the forces that have shaped the earth.

TWO MODELS of the mechanism responsible for earthquakes have been proposed in an attempt to put earthquake prediction on a sound theoretical basis. One view, called the dilatancy-diffusion model, was developed mainly in the U.S. The alternative, sometimes known as the dilatancy-instability model, was formulated in the U.S.S.R. The black-outlined curves show the expected precursory signals according to the American model; the colored curves show the expected precursory signals according to the Russian model. (Dilatancy is the technical term used to describe the inelastic increase in volume that begins when the stress on a rock reaches half the breaking strength of the rock.) The illustration is based on the work of Christopher H. Scholz, Lynn R. Sykes and Y. P. Aggarwal of the U.S. and V. I. Myachkin and G. A. Sobolev of the U.S.S.R.

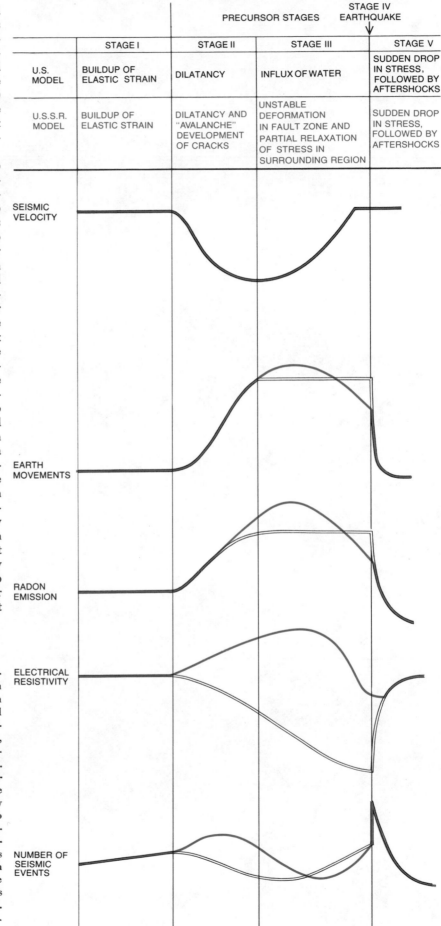

	STAGE I	STAGE II	STAGE III	STAGE IV EARTHQUAKE ↓ — STAGE V
PRECURSOR STAGES				
U.S. MODEL	BUILDUP OF ELASTIC STRAIN	DILATANCY	INFLUX OF WATER	SUDDEN DROP IN STRESS, FOLLOWED BY AFTERSHOCKS
U.S.S.R. MODEL	BUILDUP OF ELASTIC STRAIN	DILATANCY AND "AVALANCHE" DEVELOPMENT OF CRACKS	UNSTABLE DEFORMATION IN FAULT ZONE AND PARTIAL RELAXATION OF STRESS IN SURROUNDING REGION	SUDDEN DROP IN STRESS, FOLLOWED BY AFTERSHOCKS

SEISMIC VELOCITY

EARTH MOVEMENTS

RADON EMISSION

ELECTRICAL RESISTIVITY

NUMBER OF SEISMIC EVENTS

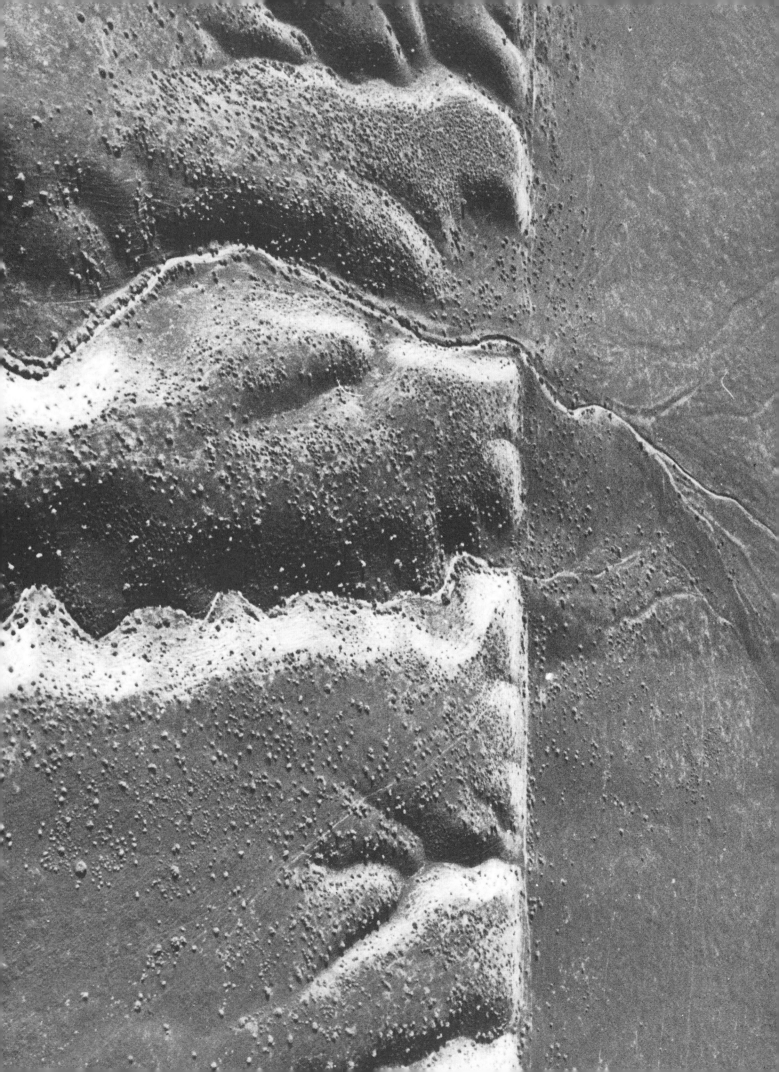

by Don L. Anderson
November 1971

This well-known break in the earth's crust is actually not one fault but a system of faults. The break separates a northward-moving wedge of California, including Los Angeles, from the rest of North America

The San Fernando earthquake that occurred at sunrise on February 9, 1971, jolted many southern Californians into acute awareness that California is earthquake country. Although it was only a moderate earthquake (6.6 on the Richter scale), it was felt in Mexico, Arizona, Nevada and as far north as Yosemite National Park, more than 250 miles from San Fernando. It was recorded at seismic stations around the world. In spite of its relatively small size the San Fernando earthquake was extremely significant because it happened near a major metropolis and because its effects were recorded on a wide variety of seismic instruments. Within hours the affected region was aswarm with geologists mapping faults and seismologists installing portable instruments to monitor aftershocks and the deformation of the ground. It was immediately clear from data telemetered to the Seismological Laboratory of the California Institute of Technology in Pasadena from the Caltech Seismic Network that the earthquake was not centered on the much feared San Andreas fault or, for that matter, on any fault geologists had labeled as active. The faults in the area, however, are all part of the San Andreas fault system that covers much of California.

The San Andreas fault system (and its attendant earthquakes) is part of a global grid of faults, chains of volcanoes

and mountains, rifts in the ocean floor and deep oceanic trenches that represent the boundaries between the huge shifting plates that make up the earth's lithosphere. The concept of moving plates is now fundamental to the theory of continental drift, which was long disputed but is now generally accepted in modified form on the basis of voluminous geological, geophysical and geochemical evidence. The theory had received strong support from the discovery that the floors of the oceans have a central rise or ridge, often with a rift along the axis, that can be traced around the globe. Within the rift new crustal material is continuously being injected from the plastic mantle below, forming a rise or ridge on each side of the rift. The newly formed crustal material slides away from the ridge axis. Since the magnetic field of the earth periodically reverses polarity, the newly injected material "freezes" in stripes parallel to the ridge axis, whose north-south polarity likewise alternates. By dating these stripes one can estimate the rate of seafloor spreading.

The San Andreas fault system forms the boundary between the North American plate and the North Pacific plate and separates the southwestern part of California from the rest of North America. In general the Pacific Ocean and that part of California to the west of the San Andreas fault are moving northwest

with respect to the rest of the continent, although the continent inland at least as far as Utah feels the effects of the interactions of these plates.

The relative motion between North America and the North Pacific has been estimated in a variety of ways. Seismic techniques yield values between 1½ and 2½ inches per year. The ages of the magnetic stripes on the ocean floor indicate a rate of about 2⅔ inches per year. Geodetic measurements in California give rates between two and three inches per year. The ages of the magnetic anomalies off the coast of California indicate that the oceanic rise came to intersect the continent at least 30 million years ago. Geologists and geophysicists at a number of institutions (notably the University of Cambridge, Princeton University and the Scripps Institution of Oceanography) have proposed that geologic processes on a continent are profoundly affected when a continental plate is intersected by an oceanic rise. At the rates given above the total displacement along the San Andreas fault amounts to at least 720 miles if motion started when the rise hit the continent and if all the relative motion was taken up on the fault. Displacements this large have not been proposed by geologists, but the critical tests would involve correlation of geology in northern California with geology on the west coast of Baja California, an area that has only recently been studied in detail. One can visualize how the west coast of North America may have looked 32 million years ago by closing up the Gulf of California and moving central and northern California back along the San Andreas fault to fit into the pocket formed by the coastline of the northern half of Baja California. This places all of California west of the San Andreas fault south of

DISPLACEMENT ALONG SAN ANDREAS FAULT is clearly visible in the aerial photograph [opposite page] of a region a few miles north of Frazier Park, Calif., itself 65 miles northwest of Pasadena, where the fault runs almost due east and west. This east-west section of the San Andreas fault is part of the "big bend," where the fault appears to be locked. The photograph is reproduced with north at the right. The hilly region to the left (south) of the fault line is moving upward (westward) with respect to the flat terrain at the right, causing clearly visible offsets in the two largest watercourses as they flow onto the alluvial plain.

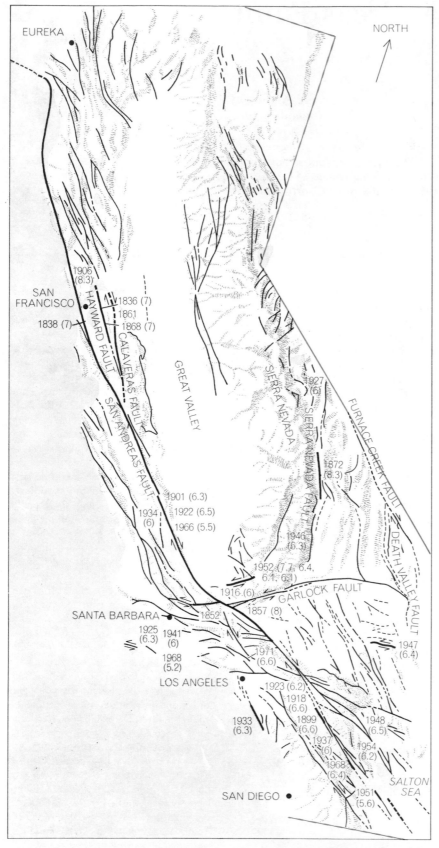

SIMPLIFIED FAULT MAP of California identifies in heavy black lines the faults that
have given rise to major earthquakes since 1836. The magnitude of all but two of the earthquakes is given in parentheses next to the year of occurrence. For events that predated
the introduction of seismological instruments the magnitudes are estimated from historical accounts. For two major events, the earthquakes of 1852 and 1861, information is too
sparse to allow a magnitude estimate. Arrows parallel to the faults show relative motion.

the present Mexican border [*see illustration on page 48*].

California is riddled with faults, most
of which trend roughly northwest-southeast, like most of the other tectonic and
geologic features of California (such as
the Sierra Nevada and the Coast
Ranges). The prominent exceptions are
the east-west-trending transverse ranges
and faults that make up a band some
100 miles wide extending inland from
between Los Angeles and Santa Barbara. The San Gabriel Mountains, which
form the rugged backdrop to Los Angeles, are part of this complex geologic
region, and it was here that the San Fernando earthquake struck. The northeast-
trending Garlock fault and the Tehachapi Mountains, which separate the Sierra
Nevada and the Mojave Desert, also
cut across the general grain of California. The area to the west of most of
the northwest-trending faults is moving
northwest with respect to the eastern
side. This is called right-lateral motion.
If one looks across the fault from either
side, the other side is moving to the
right.

Motion on the Garlock fault is left-
lateral, which, combined with the right-
lateral motion on the San Andreas fault,
means that the Mojave Desert is moving
eastward with respect to the rest of California. Parts of the faults that have been
observed or inferred to move as a result
of earthquakes in historic times are
shown in the illustration at the left. Also
shown are the dates of the earthquakes
and the magnitude of some of the more
important ones. In general both the
length of rupture and the total displacement are greater for the larger earthquakes. Horizontal displacements as
great as 21 feet were observed along the
San Andreas fault after the San Francisco earthquake of 1906, which had a
magnitude of 8.3 on the Richter scale.
(The Richter scale, devised by Charles
F. Richter of Cal Tech, is logarithmic.
Although each unit denotes a factor of
10 in ground amplitude, or displacement, the actual energy radiated by an
earthquake is subject to various modifi-
cations.) The San Fernando earthquake
produced displacements of six feet,
whose direction was almost equally divided between the horizontal and the
vertical.

The trend of the San Andreas fault
system is roughly northwest-southeast
from San Francisco to the south end of
the Great Central Valley (the San Joa-
quin Valley) and again from the north
of the Salton Sea depression to the Mexican border. The motion along the faults

in these areas is parallel to the fault and is mainly strike-slip, or horizontal. Between these two regions, from the south end of the San Bernardino Mountains to the Garlock fault, the faults bend abruptly and run nearly east and west, producing a region of overthrusting and crustal shortening [*see illustration below*]. The attempt of the southern California plate to "get around the corner" as it moves to the northwest is responsible for the complex geology in the transverse ranges, for the abrupt change in the configuration of the coastline north of Los Angeles and ultimately for the recent San Fernando earthquake. The big bend of the San Andreas fault is commonly regarded by seismologists as being locked and possibly as being the location of the next major earthquake. Much of the motion in this region, however, is being taken up by strike-slip motion along faults parallel to the San Andreas fault and by overthrusting on both

sides of the fault. The displacements associated with the larger earthquakes in southern California in the vicinity of the big bend have averaged out to about 2½ inches per year since 1800. The Kern County earthquake of 1952 (magnitude 7.7) apparently took care of most of the accumulated strain, at least at the north end of the big bend, that had built up since the Fort Tejon earthquake of 1857 (magnitude 8).

The San Andreas fault system cannot be completely understood independently of the tectonics and geology of most of the western part of North America and the northeastern part of the Pacific Ocean. This vast region is itself only a part of the global tectonic pattern, all parts of which seem to be interrelated. The earthquake, tectonic and mountain-building activities of western North America are intimately related to the relative motions of the Pacific and North American plates. Just as it is misleading

to think of the San Andreas fault as an isolated mechanical system, so it is misleading to think of the entire San Andreas fault as a single system. The part of the fault that lies in northern California was activated earlier and has moved farther than the southern California section. The northern portion is less active seismically than the southern section and seems to have been created in a different way. It is also moving in a slightly different direction.

Measuring Displacements

There are several ways to measure displacements on major faults. Fairly recent displacements are reflected in offset stream channels [*see illustration on page 40*]. Many such offsets measured in thousands of feet are apparent across the San Andreas fault in central California, some of which can be directly related to earthquakes of historic times.

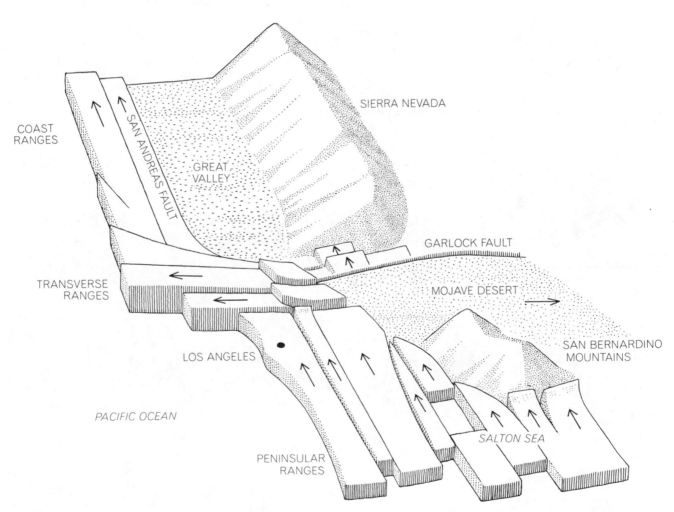

MOTION OF EARTH'S CRUST in southern California is generally northwest except where the lower group of blocks encounters the deep roots of the Sierra Nevada. At this point the blocks are diverted to the left (west), creating the transverse ranges and a big bend in the San Andreas fault system. Above the bend the blocks continue their northwesterly march, carrying the Coast Ranges with them. The Salton Sea trough at the lower right evidently represents a rift that has developed between two blocks.

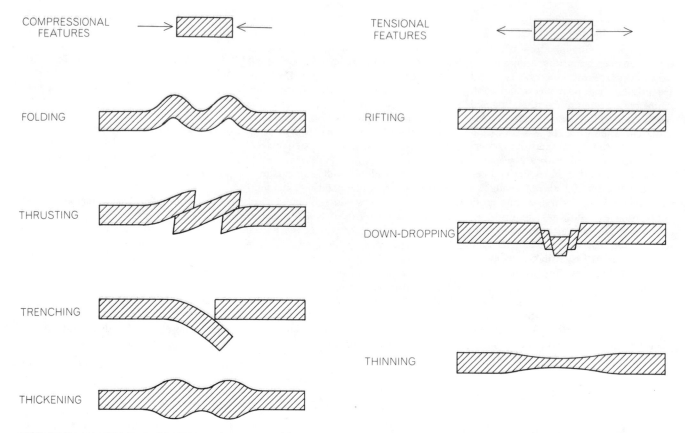

COMPRESSIONAL FEATURES

TENSIONAL FEATURES

FOLDING

THRUSTING

TRENCHING

THICKENING

RIFTING

DOWN-DROPPING

THINNING

RESPONSE OF CRUSTAL PLATES to compression (*left*) and tension (*right*) accounts for most geologic features. According to the recently developed concept of plate tectonics, the earth's mantle is covered by huge, rigid plates that can be colliding, sliding past one another or rifting apart. The rifting usually occurs in the ocean floor. The San Andreas fault marks the location where two plates are sliding past each other. Plate tectonics helps to explain how the continents have drifted into their present locations.

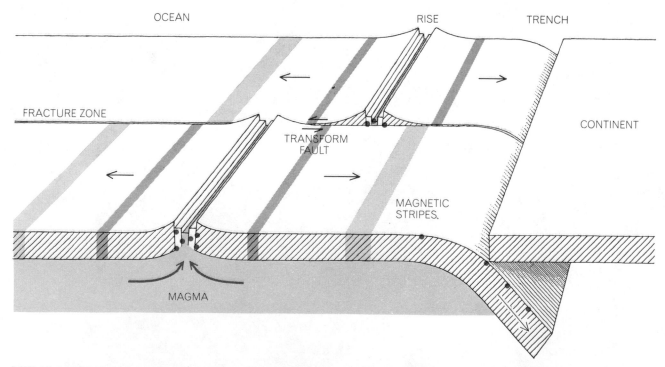

OCEAN   RISE   TRENCH

FRACTURE ZONE

TRANSFORM FAULT

CONTINENT

MAGNETIC STRIPES

MAGMA

RIFT IN OCEAN FLOOR (*color*) initiates three major features of oceanic plate tectonics. The rift is bordered by a rise or ridge created by magma pushed up from the mantle below. The magma solidifies with a magnetic polarity corresponding to that of the earth. When, at long intervals, the earth's polarity reverses, the polarity of newly formed crust reverses too, resulting in a sequence of magnetic "stripes." A trench results when an oceanic plate meets a continental plate. A fracture zone and transform fault result when two plates move past each other. Earthquakes (*dots*) accompany these tectonic processes. The earthquakes in the vicinity of a rise and along a transform fault are shallow. Deep-focus earthquakes occur where a diving oceanic plate forms a trench.

Erosion destroys this kind of evidence very quickly. By matching up distinctive rock units that have been broken up and moved with respect to each other it is possible to document offsets of tens to hundreds of miles. A sedimentary basin often holds debris that could not possibly have been derived from any of the local mountains; matching up these basins with the appropriate source region on the other side of the fault can provide evidence of still larger displacements. When these various kinds of information are combined, one obtains a rate of about half an inch per year for motion on the San Andreas fault in northern and central California over the past several tens of millions of years.

This is much less than the 2½ inches per year that is inferred for the rate of separation of Baja California and mainland Mexico and the rate that is inferred from seismological studies in southern California. There are several possible explanations for the discrepancy. Northern and southern California may be moving at different rates; this seems unlikely since they are both attached to the same Pacific plate. On the other hand, part of the compression in the transverse ranges may result from a differential motion between the two parts of the state. Another possibility is that all of the relative motion between the North American plate and the Pacific plate is not being taken up by the San Andreas fault or even by the San Andreas fault system but extends well inland. The fracture zones of the Pacific seem to affect the geology of the continent for a distance of at least several hundred miles.

The Great Central Valley and the Sierra Nevada lie between two major fracture zones that abut the California coast: the Mendocino fracture zone and the Murray fracture zone. The transverse ranges, the Mojave Desert and the Garlock fault are all in line with the Murray fracture zone. Recent volcanism lines up with the extensions of the Clarion fracture zone and the Mendocino fracture zone. The basins and range geological province of the western U.S., a region of crustal tension and much volcanism, may represent a broad zone of deformation between the Pacific plate and the North American plate proper. Seismic activity is certainly spread over a large, diffuse region of the western U.S.

Although the subject has been quite controversial, most geologists are now willing to accept large horizontal displacements on the faults in California, particularly the San Andreas. Displacements as large as 450 miles of right-lat-

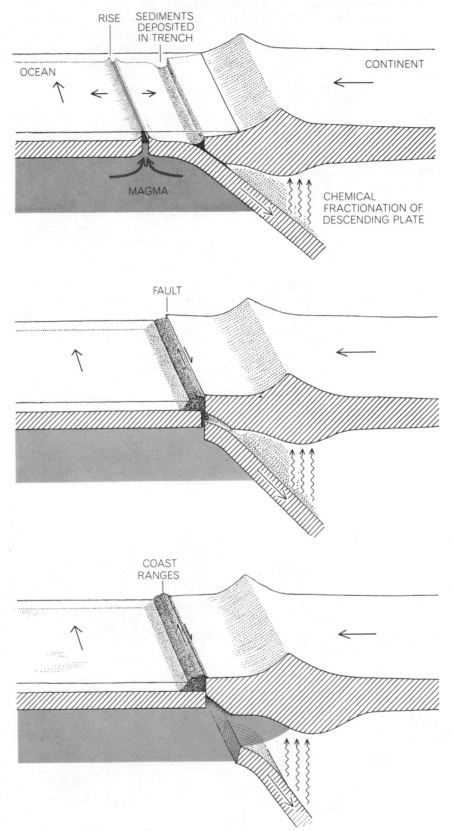

INTERACTION BETWEEN RISE AND TRENCH leads to mutual annihilation. The trench, formed as the oceanic plate dives under the continental plate, slowly fills with sediments carried by rivers and streams (*top*). Meanwhile the melting of the descending slab adds new material to the continent from below. When the axis of the rise reaches the edge of the continent, the flow of magma into the rift is cut off and trench sediments are scraped onto the western (that is, left) part of the oceanic plate (*middle*). The descending plate disappears under the continent and the sediments travel with the oceanic plate (*bottom*). The northern part of the San Andreas fault may have been formed in this way.

eral slip have been proposed for the northern segment of the fault. Displacements on the southern San Andreas fault are put at no more than 300 miles. This discrepancy has been puzzling to geologists. My own conclusion is that the part of northern and central California west of the San Andreas fault has moved northwest more than 700 miles and that the southern San Andreas fault has slipped about 300 miles, which makes the apparent discrepancy even worse. The discrepancy disappears if one drops the concept of a single San Andreas fault and admits the possibility that the two segments of the fault were initiated at different times.

The two-fault hypothesis is supported by straightforward extrapolation of the record on the ocean floor. The two San Andreas faults formed at different times, in different ways and may be moving at different rates. The record indicates that the western part of North America caught up with a section of the East Pacific rise somewhere between 25 million and 30 million years ago. Before the collision a deep oceanic trench existed off the coast such as now exists farther to the south off Central America and South America. The trench had existed for many millions of years, receiving

sediments from the continent; subsequently the sediments were carried down into the mantle by the descending oceanic plate, which was diving under the continent. Based on what we know of trench areas that are active today one can assume that the plate sank to 700 kilometers and that the process was accompanied by earthquakes with shallow, intermediate and deep foci.

The Origin of Continents

Let us examine a little more closely what happens when an oceanic rise, the source of new oceanic crust, approaches a trench, which acts as a sink, or consumer of crust. Evidently the rise and the trench annihilate each other. The oceanic crust and its load of continental debris, which was formerly in the trench, rise because the crust is no longer connected to the plate that was plunging under the continent. The trench deposits are so thick they eventually rise above sea level and become part of the continent. The deposits are still attached to the oceanic plate, however, and travel with it [see illustration on page 91].

In the case of the Pacific plate off California the deposits move northwest

with respect to the continent. This is the stuff of coastal California north of Santa Barbara, particularly the Coast Ranges. According to this view, the northern segment of the San Andreas fault was born at the same time as northern California. The rise and the trench initially interacted near San Francisco, which then was near Ensenada in Baja California. Ensenada in turn was near the northern end of the Gulf of California, which was then closed.

The tectonics and geologic history of California, and in fact much of the western U.S., are now beginning to be understood in terms of the new ideas developed in the theories of sea-floor spreading, continental drift and plate tectonics. Many of the basic concepts were laid down by the late Harry H. Hess of Princeton and Robert S. Dietz of the Environmental Science Services Administration. Tanya Atwater of the University of California at San Diego and Warren Hamilton of the U.S. Geological Survey and their colleagues have made particularly important contributions by applying the concepts of plate tectonics to continental geology. We now know that the outer layer of the earth is immensely mobile. This layer, the lithosphere, is relatively cold and

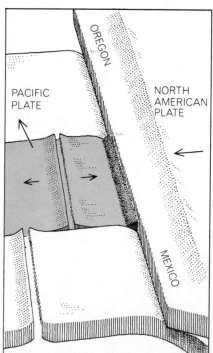
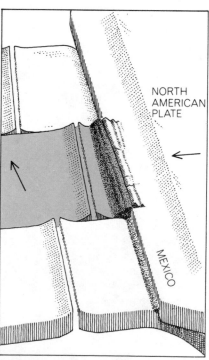
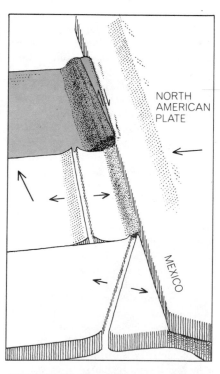

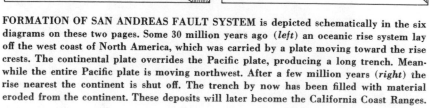

FORMATION OF SAN ANDREAS FAULT SYSTEM is depicted schematically in the six diagrams on these two pages. Some 30 million years ago (*left*) an oceanic rise system lay off the west coast of North America, which was carried by a plate moving toward the rise crests. The continental plate overrides the Pacific plate, producing a long trench. Meanwhile the entire Pacific plate is moving northwest. After a few million years (*right*) the rise nearest the continent is shut off. The trench by now has been filled with material eroded from the continent. These deposits will later become the California Coast Ranges.

NORTHERN SECTION of San Andreas fault is created when the former trench deposits become attached to the northward-moving Pacific plate (*left*). The San Andreas fault lies between the two opposed arrows indicating relative plate motions. Meanwhile to the south a tilted rise crest

rigid and slides around with little resistance on the hot, partially molten asthenosphere.

Where the crust is thick, as it is in continental regions, the temperatures become high enough in the crust itself to cause certain types of crustal rocks to lose their strength and to offer little resistance to sliding. There is thus the possibility that the upper crust can slide over the lower crust and that the moving plate can be much thinner than is commonly assumed in plate-tectonic theory. The molten fraction of the asthenosphere, called magma, rises to the surface at zones of tension such as the mid-oceanic rifts to freeze and form new oceanic crust. The new crust is exposed to the same tensional forces (presumably gravitational) that caused the rift in the first place; therefore it rifts in turn and subsequently slides away from the axis of the rise. In addition to providing the magma for the formation of new crust, the melt in the asthenosphere serves to lubricate the boundary between the lithosphere and the asthenosphere and effectively decouples the two. The rise is one of the types of boundary that exist between lithospheric plates and is the site of small, shallow tensional earthquakes.

When two thin oceanic lithospheric plates collide, one tends to ride over the other, the bottom plate being pushed into the hot asthenosphere. The boundary becomes a trench. When the lower plate starts to melt, it yields a low-density magma that rises to become part of the upper plate; this magma becomes the rock andesite, which builds an island arc on what is to become the landward, or continental, side of the trench. (The rock takes its name from the Andes of South America. Mount Shasta in California is primarily andesite, as are the island arcs behind the trenches that surround the Pacific.) The thickness of the crust is essentially doubled as a result of the underthrusting. The material remaining in the lower plate is now denser than the surrounding material in the asthenosphere, both because it has lost a low-density fraction and because it is colder; thus it sinks farther into the mantle. In some parts of the world the downgoing slab can be tracked by seismic means to 700 kilometers, where it seems to bottom out. By this process new light material is added to the crust and new dense material is added to the lower mantle. A large part of what comes up stays up; a large part of what goes down stays down.

The introduction of chemical fractionation and a mechanism for "unmixing" makes the process different from the one customarily visualized, in which gigantic convection cells carry essentially the same material in a continuous cycle. The new process is able to explain in a convincing way how continents are formed and thickened. As the continent thickens and rises higher because of buoyancy, erosional forces become more effective and dump large volumes of continental sediments into the coastal trenches. A portion of the sediments is ultimately dragged under the continent to melt and form granite. The light granitic magma rises to form huge granitic batholiths such as the Sierra Nevada. A batholith is a large mass of granitic rock formed when magma cools slowly at great depth in the earth's crust. It is carried to the surface by uplifting forces and exposed by erosion.

The concept of rigid plates moving around on the earth's surface and interacting at their boundaries has been remarkably successful in explaining the evolution of oceanic geology and tectonics. The oceanic plates seem to behave rather simply. Tension results in a rise, compression results in a trench and lateral motion results in a transform fault

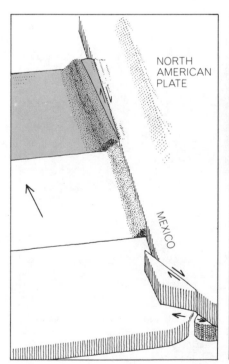
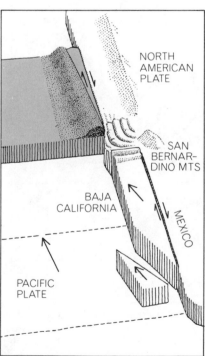
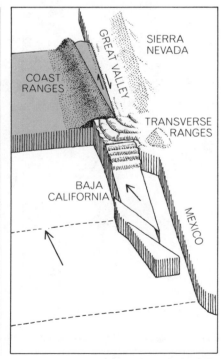

(not yet visible in the first pair of diagrams) is ready to encounter the continent end on at a break in the coastline south of Baja California. The collision (right) breaks off a part of the Baja California peninsula, which becomes attached to the Pacific plate and starts its journey to the northwest.

SOUTHERN SECTION of San Andreas fault is now fully activated (left) as the Baja California block begins sliding past the North American plate and collides with deeply rooted structures to the north, the Sierra Nevada and San Bernardino Mountains, which deflect the block to the west. More of Baja California breaks loose, opening up the Gulf of California. As Baja California continues to move northwestward (right) the Gulf of California steadily widens. The compression at the north end of the Baja California block creates the transverse ranges, which extend inland from the vicinity of present-day Santa Barbara.

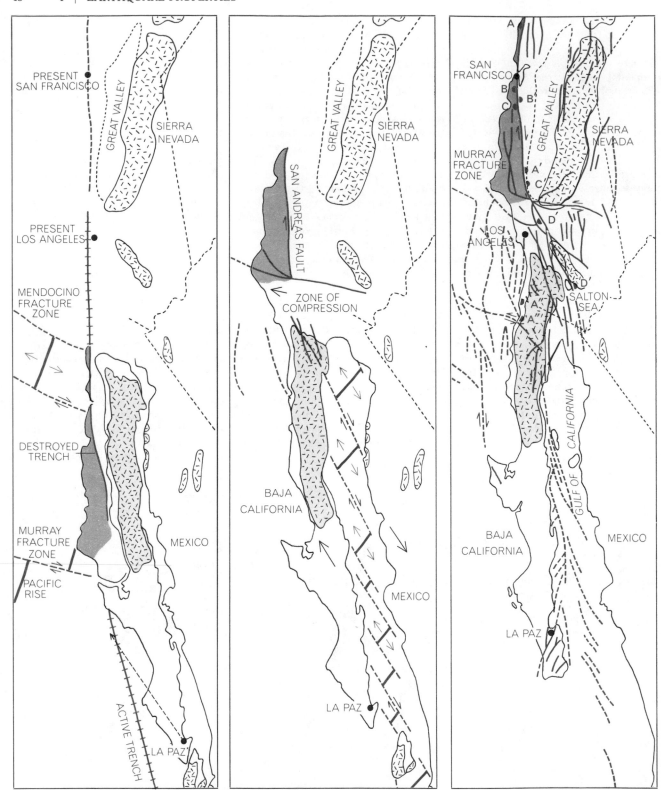

EARLY AND LATE STAGES in the history of the San Andreas fault are depicted. Twenty-five million years ago (*left*) Baja California presumably nestled against mainland Mexico. The first section of oceanic rise between the Murray fracture zone and the Pioneer fracture zone has just collided with the continent. Trench deposits are uplifted and become part of the Coast Ranges of California. The block containing the present San Francisco area (*dark color*) is about to start its long northward journey. A block immediately to the east (*light color*) becomes attached to the Pacific plate and eventually is jammed against the San Bernardino Mountains. Three million years ago (*middle*) the Gulf of California has started to open. As the peninsula moves away from mainland Mexico a series of rifts appear, fill with magma and are offset by numerous fractures. Baja California may have been torn off in one piece or in slivers. Southwest California and Baja California today (*right*) continue to slide northwest against the North American mainland. The illustration shows major fault systems and offshore fracture zones. On the basis of unique rock formations geologists infer that the Los Angeles area has moved northwest about 130 miles (*D' to D*) in the past 20 million years or less. Other studies indicate that the Palo Alto region has been carried about 200 miles (*C' to C*). Coastal rocks to the north of San Francisco have been displaced at least 300 miles (*A' to A*) and perhaps as much as 650 miles (*A'' to A*) in the past 30 million years.

and a fracture zone [*see bottom illustration on page 44*]. The interaction of oceanic and continental plates or of two continental plates is apparently much more complicated, and this is one reason the new concepts were developed by study of the ocean bottom rather than continental geology.

The boundary between two oceanic plates can be a deep oceanic trench, an oceanic rise or a strike-slip fault depending on whether the plates are approaching, receding from or moving past each other. The forces involved are respectively compressional, tensional and shearing. When a thick continental plate is involved, compression can also result in high upthrust and folded mountain ranges. The Himalayas resulted from the collision of the subcontinent of India with Asia. I shall show that the transverse ranges in California were formed in a similar way. Tension can result in a wide zone of crustal thinning, normal faulting and volcanism; it can also create a fairly narrow rift of the kind found in the Gulf of California and the Red Sea [*see top illustration on page 44*].

The interaction of western North America with the Pacific plate has led to large horizontal motions along the San Andreas fault, to concentrated rifting as in the Salton Sea trough and the Gulf of California, to diffuse rifting and normal faulting accompanied by volcanism in the basins and range province of California, Nevada, Utah and Arizona, to large vertical uplift by overthrusting as in the transverse ranges north and west of Los Angeles, to the generation of large batholiths such as the Sierra Nevada and to the incorporation of deep-sea trench material on the edge of the continent. Ultimately the geology, tectonics and seismicity of California can be related to the collision of North America with the Pacific plate.

Most of the Pacific Ocean is bounded by trenches and island arcs. Trenches border Japan, Alaska, Central America, South America and New Zealand. Island arcs are represented by the Aleutians, the Kuriles, the Marianas, New Guinea, the Tongas and Fiji. The arcs are themselves bordered by trenches. All these areas are characterized by andesitic volcanism and deep-focus earthquakes. Western North America is lacking a trench and has only shallow earthquakes, but the geology indicates that there was once a trench off the West Coast, and in fact there was once a rise. The present absence of a rise and a trench, the absence of deep-focus earthquakes and the existence of uplifted deep-sea sediments are all related.

Tracing back the history of the interaction of the Pacific plate with the North American plate, one is forced to conclude that the northern part and the southern part of the San Andreas fault originated at different times and in different ways. The northern part was evidently formed about 30 million years ago when a portion of rise between the Mendocino fracture zone and the Murray fracture zone approached an offshore trench bordering the southern part of North America. At that time the west coast of North America resembled the present Pacific coast of South America: there was a deep trench offshore, high mountains paralleled the coastline and large underthrusting earthquakes were associated with the downgoing lithosphere.

Origin of the Fault

As the rise approached the continent both the geometry and the dynamics of interaction changed [*see illustrations on pages 46 and 47*]. Depending on the spreading rate of the new crust generated at the rise and the rate at which the rise itself approaches the continent, the relative motion between the rise and the continental plate will decrease, stop or reverse when the rise hits the trench. The forces keeping the trench in existence will therefore decrease, stop or reverse, leading to uplift of the sedimentary material that has been deposited in the trench. In classical geologic terms these are known as eugeosynclinal deposits. Although they have been exposed to only moderate temperatures, they have been subjected to great pressures, both hydrostatic (owing to their depth of burial) and directional (owing to the horizontal compressive forces between the impinging plates). Eugeosynclinal sediments are therefore strongly deformed and become even more so as they are contorted and sheared during uplift. Much of the western edge of California and Baja California is underlain by this material, called the Franciscan formation. The formation is physically attached to the Pacific plate and is therefore moving northwest with respect to the rest of North America. The present boundary is the northern part of the San Andreas fault. Today this section of the San Andreas system extends from Cape Mendocino, north of San Francisco, to somewhere south and east of Santa Barbara, near the beginning of the great bend of the San Andreas fault, where the San Andreas and the Garlock faults intersect.

Meanwhile, 30 million years ago, another section of the rise south of the Murray fracture zone was still offshore, together with an active trench. Baja California was still attached to the mainland of Mexico and the Gulf of California had not yet opened up. The southern part of the San Andreas fault had not yet been formed.

The abrupt change in the direction of the coastline south of the tip of Baja California suggests that here the rise approached the continent more end on than broadside. A sliver of existing continent was welded onto the Pacific plate and rifted away from Mexico, thus forming Baja California and the Gulf of California. Thereafter Baja California participated in the northwesterly motion of the Pacific plate, with the result that the Gulf of California widened progressively with time.

About five million years were required for northern California, which had broken off from Baja California, to be carried about 200 miles to the northwest. At the end of that time the Gulf of California and the Salton Sea trough had not yet opened. The faults that delineate the major geologic blocks in southern California had not yet been activated. The block bearing the San Gabriel fault, now north of San Fernando, occupied the future Salton Sea trough. The transverse ranges will eventually be formed from the Santa Barbara, San Gabriel and San Bernardino blocks by strong compression from the south when Baja California breaks loose from mainland Mexico. This also opens up the Gulf of California and the Salton Sea trough.

As northern California is being carried away from Baja California by the Pacific plate another segment of oceanic rise south of the Murray fracture zone approaches the southern half of Baja California, where the situation described above is repeated except that the rise crest encounters a sharp bend in the coastline and the trench hits just south of the tip of the peninsula. Now instead of approaching the continent more or less broadside the rise approaches the continent end on. Mainland Mexico is still decoupled from that part of the Pacific plate to the west of the rise by the rise and the trench. Baja California, however, is now coupled to the northwestward-moving Pacific plate and Baja California is torn away from the mainland. This happened between four and six million years ago. Magma from the upper mantle wells up into the rift, forming a new rise that works its way north into the widening gulf. Alternatively, the entire peninsula of Baja California could have broken off from the mainland at the

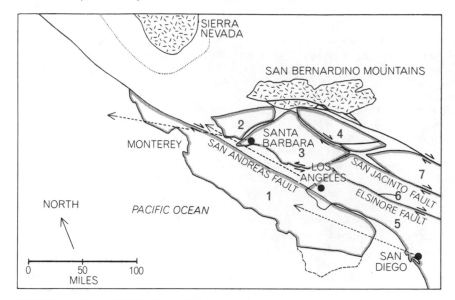

SEQUENCE OF SIMPLIFIED VIEWS shows the movement of major blocks in southern California over the past 12 million years. In the first view (*above*) the Gulf of California has not yet appreciably opened but the block carrying the Coast Ranges (*1*) has started to move rapidly northwest with activation of northern portion of San Andreas fault. Dots show origin and arrows show displacement of San Diego, Los Angeles and Santa Barbara.

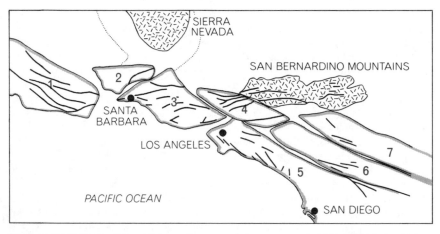

TWO MILLION YEARS AFTER ACTIVATION of the southern portion of the San Andreas fault four blocks (*2, 3, 4, 7*) have been forced against the deep roots of the Sierra Nevada and San Bernardino Mountains. Compressive forces create the transverse ranges. Meanwhile the block carrying the Coast Ranges (*1*) has been carried far to the northwest.

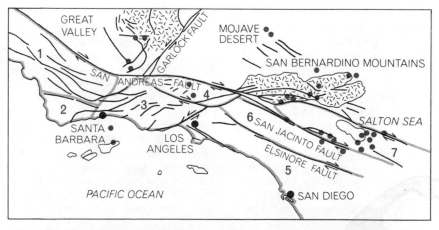

GEOLOGY OF SOUTHERN CALIFORNIA TODAY is dominated by compressive forces operating in the big bend of the San Andreas fault, which connects the southern and northern parts of the system. Colored dots show the location of earthquakes in the recent past.

same time. As the peninsula, including parts of southern California, moves north it collides with parts of the continent that are still attached to the main North American plate. This results in compression, overthrusting and shearing and the eventual formation of the transverse ranges.

The southern part of the San Andreas fault system was therefore formed by the rifting off of a piece of continent. Today it represents the boundary between two parts of the continental plate that are moving with respect to each other. This part of the San Andreas fault was formed well east, or inland, of the southward projection of the northern San Andreas.

The northerly march of southern California and Baja California seems to have been blocked when the moving plate encountered the thick continental crust to the north, particularly the massive granitic San Bernardino mountain range, which includes the 11,485-foot San Gorgonio Mountain. Since large and high mountain ranges have deep roots, the crust in this region is probably much thicker than normal, perhaps as thick as 50 kilometers. Earthquakes in this region are all shallower than 20 kilometers, which may be the thickness of the sliding plates. The blocks veer westward and are strongly overthrust as they attempt to get around the obstacle; this movement generates the big bend in the San Andreas fault system. The deflected blocks eventually join up with the northern California block.

Earthquake Country

From a social and economic point of view earthquakes are one of the most important manifestations of plate interaction. From a scientific point of view they supply a third dimension to the study of faults and the nature of the interactions between crustal blocks, including the stresses involved and the nature of the motions.

Seismologists at the University of California at Berkeley and at the Cal Tech Seismological Laboratory have been keeping track of earthquake activity in California for more than 40 years. Both groups have installed arrays of seismometers that telemeter seismic data to their laboratories for processing and dissemination to the appropriate public agencies. During the 36-year period 1934 through 1969 there were more than 7,300 earthquakes with a Richter magnitude of 4 or greater in southern California and adjacent regions [*see illustration on page 52*]. Many thousands more earthquakes of smaller magnitude are routinely located and reported in the seismological bulletins. Although damage depends on local geological conditions and the nature of the earthquake, a rough rule of thumb is that a nearby earthquake of magnitude 3.5 or greater can cause structural damage. The average annual number of earthquakes of magnitude 3 or greater in southern California recorded since 1934 is 210; the number in any one year has varied from a low of 97 to a high of 391. The strongest earthquake in this period was the Kern County event of magnitude 7.7 in 1952. The aftershocks of that event increased the total number of events for several years thereafter.

In general the larger the earthquake, the greater the displacement across a fault and the greater the length of fault that breaks. The great earthquakes of 1906 and 1857 respectively caused large displacements across the northern and central parts of the San Andreas fault and relieved the accumulated strain in these areas. The accumulation of strain in southern California is relieved mainly by slip on a series of parallel faults and by overthrusting on faults at an angle to the main San Andreas system; that is what happened in the Kern County and San Fernando earthquakes. The unique east-west-trending transverse ranges were formed in this way. In the process deep-seated ancient rocks were uplifted and exposed by erosion.

Another seismically active area associated with major faults is south of the Mojave Desert near San Bernardino, where the faults show a sudden change in direction. The central part of the Mojave Desert is also moderately active. This is consistent with the idea that the sliding lithosphere is diverted by the San Bernardino Mountains. Faults and evidence of relatively recent volcanic events abound in the area. The northern part of Baja California is also quite active. An interesting feature of seismicity maps of southern California is the alignment of earthquakes in zones that trend roughly northeast-southwest, approximately at right angles to the major trend of the San Andreas system.

The map on the next page shows that the San Andreas fault itself has played only a small role in the seismicity of southern California over the past 30-odd years. One must not forget, however, that the great earthquake of 1857 probably broke the San Andreas fault for about 100 miles northwest and southeast from the epicenter. That epicenter is thought to have been near Fort Tejon, which is close to the projected intersection of the Garlock and San Andreas

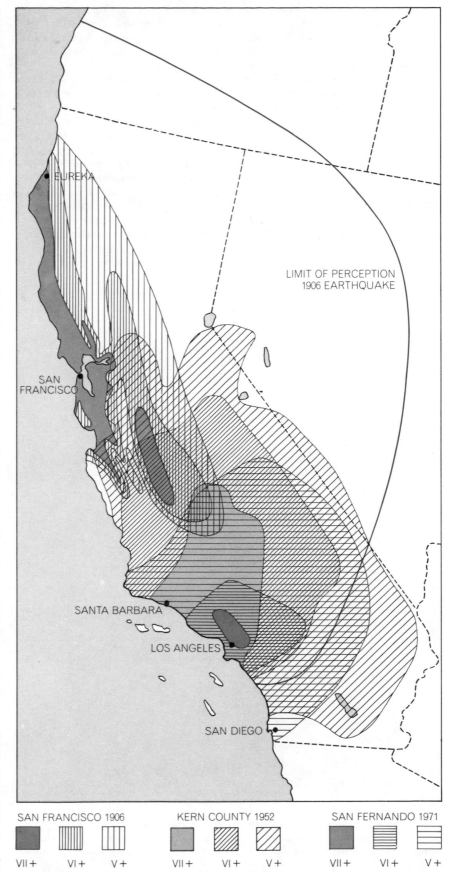

ISOSEISMAL CONTOUR MAP shows the pattern and intensity of ground-shaking produced by the 1906 San Francisco earthquake of magnitude 8.3, the 1952 Kern County earthquake of magnitude 7.7 and the 1971 San Fernando earthquake of magnitude 6.6. The Roman numerals indicate levels of perceived intensities as defined by the modified Mercalli scale. A short description of each level in the scale appears in the text on page 100.

faults; the actual location of the epicenter is uncertain by hundreds of miles because there were no seismic instruments in those days. Since that time this part of southern California has been remarkably quiet and seems to be locked, generating neither earthquakes nor creep. Activity along the San Andreas fault picks up near Coalinga, which is about midway between Bakersfield and San Francisco. Alignments of earthquakes are apparent along the San Jacinto and Imperial faults in the Salton Sea trough near the Mexican border. Although these faults lie west of the main San Andreas fault, they are part of the San Andreas system. The White Wolf fault, which is northwest of and parallel to the Garlock fault, has also been quite active, particularly after

the Kern County earthquake, which occurred on this fault. The White Wolf fault lines up with the Santa Barbara Channel area, which has similarly been quite active.

One way to quantify the seismicity of southern California is to count the number of earthquakes per year per 1,000 square kilometers and compare this figure for the world as a whole. For example, southern California averages one earthquake of magnitude 3 or greater per year per 1,000 square kilometers. Thus within the entire region there are about 200 such earthquakes per year. The rate for earthquakes of magnitude 6.6, the size of the San Fernando earthquake, is about one every five or six years. The actual rates, however, vary

considerably from year to year and depend somewhat on the time interval of the sample. The number of earthquakes decreases rapidly with size, and the average recurrence interval is not well established for the larger earthquakes. Southern California is about 10 times more actively seismically than the world as a whole, which is simply to say that California is earthquake country.

Although certain areas in southern California are relatively free of earthquakes, none is immune from their effects. One of the largest quiet areas is the western part of the Mojave Desert wedge. This is surprising because the region is bounded on the northwest and southwest by areas that are obviously under large compression, as is shown by the upthrust mountains in the transverse and Tehachapi ranges and the large overthrust earthquakes that occurred in Kern County and San Fernando. It appears that the region is being protected from the northwesterly march of the southern California–Baja California block by the San Bernardino batholith and may represent a stagnation area in the lee of the mountains. Only a small number of earthquakes are centered near San Diego, although the larger earthquakes in northern Baja California and in the mountains between San Diego and the Salton Sea are felt in San Diego. The Great Central Valley north of Bakersfield and the eastern part of the Sierra Nevada are fairly inactive, as is a large area north of Santa Barbara in the Coast Ranges.

Magnitude and Intensity

It is somewhat deceptive to plot earthquakes as small points on a map. The points represent the epicenter: the point on the surface above the initial break. Once the break is started it can continue, if the earthquake is a major one, for hundreds of miles. Earthquakes of the thrust type, which result from a failure in compression, typically first break many miles below the surface; the surface break and maximum damage can be 10 miles or more from the epicenter. The distance over which strong shocks were felt during three large California earthquakes in this century (1906, 1952 and 1971) can be represented by plotting isoseismals: lines of equal intensity [*see illustration on facing page*]. The shape of the pattern varies with the type of earthquake and with the nature of the local geology; structures on deep sedimentary basins or on uncompacted fill get a more intense shaking than structures on bedrock. The isoseismals of the

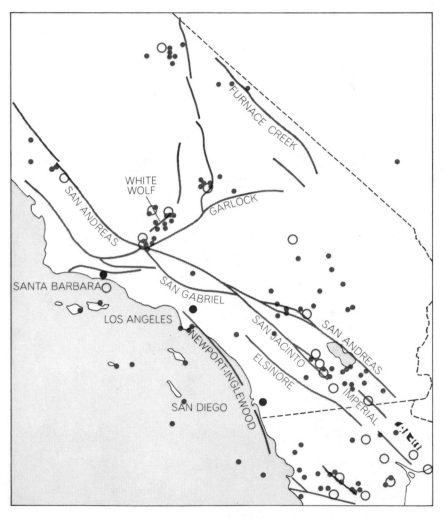

THIRTY-SIX-YEAR EARTHQUAKE RECORD shows the epicenters of all events of magnitude 5 or greater recorded in southern California and in the northern part of Baja California from 1934 through 1969. The epicenter is the point on the earth's surface above the initial break. Dots show earthquakes between 5 and 5.9 in magnitude. Open circles indicate earthquakes of magnitude 6 or greater. The hypocenter, the point of the initial break in the earth's crust, is often many miles below the surface in thrust-type earthquakes, a type frequently observed in this region. In the 36-year period southern California and adjacent regions experienced more than 7,300 earthquakes with a magnitude of 4 or more. Earthquakes are about 10 times more frequent in this area than they are in the world as a whole.

San Francisco earthquake are long and narrow, both because of the orientation of the fault and the length of the faulting and because of the northwest-southeast trend of the valleys. The orientation of the valleys in turn is controlled by the orientation of the San Andreas fault.

The public and the news media are confused about the various measures of the size of an earthquake. There are many parameters associated with an earthquake; they are usually regarded as fault parameters. They include the length, depth and orientation of the fault, the direction of motion, the rupture velocity, the radiated energy, the causal stresses and their orientation, the stress drop (which is related to the strength or the friction along the fault), the energy spectrum, the amount of offset or displacement and the time history of the motion. Most of these parameters can be estimated from seismic records, even from signals recorded several thousand kilometers from the earthquake. To obtain high precision, however, one needs records from many well-distributed seismic stations together with field observations at the site of the earthquake.

The magnitude on the Richter scale is a number assigned to an earthquake from instrumental readings of the amplitude of the seismic waves on a standard seismometer, the Wood-Anderson torsion seismometer. The amplitude must be suitably corrected for spreading and attenuation in the earth, and for instrumental response if a non-standard instrument is used. The magnitude is closely related to the energy of the earthquake, the single most important quantity by which earthquakes can be ranked one against another. If all the corrections are adequately made, a seismologist anywhere in the world will assign the same magnitude. In practice, because of the complicated radiation pattern of earthquakes and because of the distortion of the waves traveling through the earth, the initial magnitude assigned by various observatories may differ slightly. The magnitude scale is logarithmic and is open-ended at both ends. It is not a scale with a maximum value of 10, as is often reported in the press, and negative magnitudes are routinely measured by seismologists working on microearthquakes.

The intensity scale was developed for engineering purposes and is a qualitative measure of the intensity of ground vibration and structural damage. These qualitative assessments are assigned Roman numerals from I to XII. Unlike the magnitude of an earthquake, the inten-

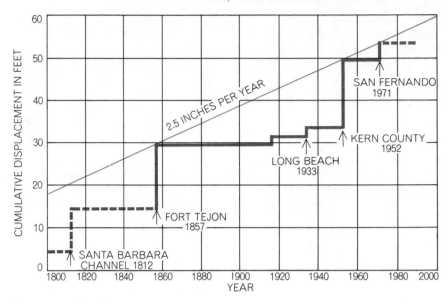

CUMULATIVE DISPLACEMENTS directly related to earthquakes indicate that southern California west of the San Andreas fault system is sliding northwestward at an average rate of 2½ inches per year. Major earthquakes relieve stresses that have built up over decades.

sity varies with distance and depends on the nature of the local ground. In general alluvial valleys, soft sediments and areas of uncompacted fill will magnify ground-shaking and will register higher intensities than adjacent areas on solid rock.

The intensity scale in common usage today is the Modified Mercalli Intensity Scale. The following characterizations of intensity, abridged from longer descriptions, indicate the kind of observations on which the Mercalli scale is based:

I. Not felt except by a very few under special circumstances. Birds and animals are uneasy; trees sway; doors and chandeliers may swing slowly.

II. Felt only by a few persons at rest, particularly on the upper floors of buildings.

III. Felt indoors, but many people do not recognize as an earthquake. Vibrations like the passing of light trucks. Duration of the shaking can be estimated.

IV. Windows, dishes and doors rattle. Walls make creaking sounds. Sensation like the passing of heavy trucks. Felt indoors by many, outdoors by few.

V. Felt by nearly everyone; many awakened. Small unstable objects are displaced or upset; plaster may crack.

VI. Felt by all; many are frightened and run outdoors. Some heavy furniture is moved; books are knocked off shelves and pictures off walls. Small church and school bells ring. Occasional damage to chimneys, otherwise structural damage is slight.

VII. Most people run outdoors. Difficult to stand up. Noticed by drivers of automobiles. Damage is negligible in

buildings of good design and construction, slight to moderate in well-built ordinary structures, considerable in poorly built or badly designed structures. Waves on ponds and pools.

Intensity VII corresponds to the general experience within five or 10 miles of the surface faults associated with the San Fernando earthquake of last February. The following intensity levels were experienced in a small area of the northern San Fernando Valley and would be widely experienced in more severe earthquakes.

VIII. Steering of automobiles affected. Frame houses move on foundations if not bolted down; loose panel walls are thrown out. Some masonry walls fall. Chimneys twist or fall. Damage is slight in specially designed structures, great in poorly constructed buildings. Heavy furniture is overturned.

IX. General panic. Damage is considerable in specially designed structures; partial collapse of substantial buildings. Serious damage to reservoirs and underground pipes. Conspicuous cracks in the ground.

X. Most masonry and frame structures are destroyed with their foundations. Some well-built wooden structures are destroyed. Rails are bent slightly. Large landslides.

XI. Few, if any, masonry structures remain standing. Bridges are destroyed. Broad fissures in the ground. Rails are bent severely.

XII. Damage is nearly total. Objects are thrown into the air.

It is clear that the Mercalli intensity scale is people-oriented; anyone can estimate the intensity from his own experience during an earthquake. The National Oceanic and Atmospheric Administration compiles information on intensities by mailing out questionnaires to a sample of the population living in an area that has experienced a sizable earthquake.

In order to obtain more exact information about the ground motions involved in earthquakes engineers have developed strong-motion accelerometers that automatically trigger and start to record when shaken severely. Most of these instruments are installed in the seismic areas of the U.S., with a particularly heavy concentration in and around Los Angeles. The instruments are expensive and must be located very close to an earthquake to provide useful data. More than 250 of the instruments were triggered during the San Fernando earthquake, and a wealth of engineering data will be provided by these records.

A strong-motion instrument records ground acceleration as a function of time. Accelerations are commonly reported as fractions of a g, the accelera-

tion due to gravity at the earth's surface. One g is roughly 10 meters per second per second. In designing a building to withstand moderate earthquakes, engineers are concerned chiefly with the maximum accelerations, the period or frequency of shaking and the duration of shaking. Buildings in earthquake-hazard regions with stringent building codes are usually designed to withstand at least .1 g of acceleration; this corresponds to an intensity of about VII on the Mercalli scale.

Although there is no direct correlation between intensity and magnitude, the zone of destruction increases as the magnitude increases for shallow-focus earthquakes. In general the larger the magnitude of an earthquake, the longer the fault length, the larger the displacement across the fault and the longer the duration of shaking. The longer fault length alone accounts for much of the increased area of destruction. For example, the San Francisco earthquake of 1906 had an intensity of VII or greater out to a distance of 500 miles from the epicenter, and this may not have been the largest California earthquake in his-

toric times. The San Francisco earthquake had a magnitude of 8.3. The 1952 Kern County earthquake (magnitude 7.7) had an intensity of VII or greater out to 50 miles. The recent San Fernando earthquake (magnitude 6.6) damaged older structures out to 25 miles. An earthquake of magnitude 5.5, the Parkfield earthquake of 1966, produced comparable damage to a distance of 10 miles.

The February Earthquake

The San Fernando earthquake occurred in the San Gabriel Mountains just north of the San Fernando Valley, a densely populated northern suburb of Los Angeles. The San Gabriel Mountains are part of the structural province of the transverse ranges: the band of east-west-trending mountains, valleys and faults that is characterized by strong and geologically recent tectonic deformation. Geologists recognize that the region is one of recent crustal shortening caused by north-south compression. The mountains, produced by buckling and thrusting, are one result of this crustal shortening. They have been thrusting over the valleys to the south for at least five million years along fault planes that dip to the north or northeast.

Although many faults are known to have been active in this area in the past several thousand years, the San Fernando earthquake produced the first historic example of surface faulting. The San Gabriel Mountains rise abruptly some 5,000 feet above the San Fernando Valley and the Los Angeles basin to the south. During the earthquake of February 9 a wedge-shaped prism of the crystalline basement rock comprising the San Gabriel Mountains was thrust over the San Fernando Valley to the southwest, thereby raising the elevation of a section of the San Gabriel Mountains and sliding it slightly to the west. The displacement is consistent with the motions that have been occurring for millions of years, as one can infer from geologic offsets and uplifts. It also agrees with the general picture presented here, namely that the transverse ranges were formed by the collision of the southern and Baja California block with the central and northern California block, and with the concept that the southern California block is being diverted to the west by the massive San Bernardino batholith. One can infer that the thickening of the crust involved in the overthrusting and uplift of the San Gabriel Mountains made this region an additional obstacle to the northwesterly march of

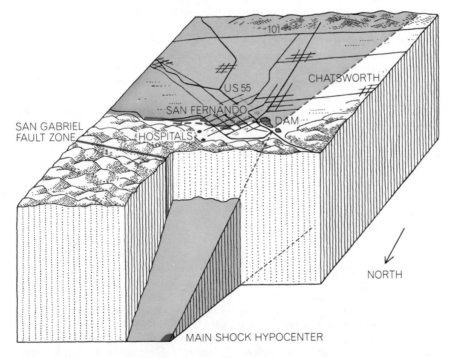

HYPOCENTER OF SAN FERNANDO EARTHQUAKE (*dark color*) of last February was 13 kilometers deep and 12 kilometers north of the area where the principal ground-shaking occurred. The earthquake collapsed sections of two hospitals in the San Fernando Valley, taking 64 lives, and so seriously weakened the earthen wall of the Van Norman Dam at the northern end of the San Fernando Valley that 80,000 people living below the dam had to leave their homes until the water level in the reservoir could be lowered. Total damage caused by the earthquake is estimated at $500 million to $1 billion. This three-dimensional view is based on a drawing prepared by two of the author's colleagues, Bernard Minster and Thomas Jordan, who worked with information supplied by geologists and geophysicists of the California Institute of Technology. The view is looking toward Los Angeles.

the southern California block. If it did, this would lend additional support to the notion that the plates in California are only 15 to 20 kilometers thick. An intriguing possibility is that the upper part of the crust is sliding with relatively little friction on a layer of rock rich in the mineral serpentine.

The hypocenter, or point of initial rupture, of the San Fernando earthquake was at a depth of 13 kilometers under the San Gabriel Mountains. The fault motion was propagated to the surface along a fault inclined northward at an angle of 45 degrees and broke the surface near the cities of San Fernando and Sylmar, at the boundary between the crystalline rocks of the mountains and the sediments of the valley [see *illustration on page 54*]. Two heavily damaged hospitals were between the epicenter and the surface break and were therefore on the upthrust, or elevated, block. The hundreds of aftershocks following the earthquake covered an area of approximately 300 square kilometers; the total volume of rock lifted up was about 2,500 cubic kilometers.

Even though the elevation difference between the peaks of the San Gabriel Mountains, such as Mount Wilson and Mount Baldy, and the floors of the adjacent valleys is impressive, it does not represent the total uplift. Erosion removes material from the mountains and deposits it in the valleys. The total amount of differential vertical motion probably exceeds two and a half miles, and horizontal displacements in the transverse ranges probably exceed 25 miles. Many thousands of earthquakes of the San Fernando type must have occurred in the area over the past several million years.

Seismic surveillance of the region with instruments dates back only four decades. In this period the northern San Fernando Valley was less active seismically than many other parts of the greater Los Angeles area, although it was comparable to the average for all southern California. On the basis of the seismic data there was no reason to believe the San Fernando area was any more or less likely than any other region of recent mountain-building in southern California to experience a large earthquake. On the other hand, the thrusting and bending associated with the geologic processes in the region, and the tilting that was associated with the earthquake and its aftershocks, suggest that a dense network of tiltmeters could provide a warning of the next large earthquake here.

II EARTHQUAKES AND EARTH STRUCTURE

INTRODUCTION

In the last section, the properties of earthquakes, their sources, and their effects on society and earthquake engineering were discussed. In this section, their great value as tools for the exploration of the earth's deep interior is demonstrated. The first of the three articles deals with a region of the earth that is profoundly disarranged by tectonic forces. It contains the world's highest mountains and some of the world's most inhospitable deserts. A study of this region of Asia and its evolution brings out the role of present earthquakes in investigation of tectonic mysteries, so that regional seismicity can be explained by a broader model. The remaining two articles are nicely complementary in discussing the way the faint echoes of an earthquake that have penetrated the earth's deep interior can be used as probes in studying the variations in interior structure. Unlike chemistry or physics, where experiments can be carefully planned and repeated under the control of the experimenter, geological studies of the earth's interior rely on sporadic surface readings, from which inferences are made about the physical properties at greater depths.

Q: What are the main shells of the earth's interior?

A: The outermost shell of the earth is called the *crust;* it is about 5 km thick under the oceans. The crust is part of the earth's *lithosphere*, a strong layer believed to be about 80 km thick that moves laterally over most of the earth's surface. Below this is the *asthenosphere*, which extends to a depth of several hundred kilometers. Beneath these outer shells is the earth's solid rocky *mantle*, which extends to a depth of 2,900 km. Below this depth is the earth's *core*, which is made up of two main parts, a liquid outer core and a solid inner core.

Q: To what basic cause can be attributed the great geological and topographical diversity of the Himalayas in Tibet and western China?

A: When the whole region is viewed from a great distance—as seen, for example, on satellite photographs—the geological complexities seem to fall into a pattern that is attributable to a geological collision between the Indian subcontinent and the rest of Eurasia. This collision began 40 million years ago and is still in progress.

Q: How far north has the Indian subcontinent moved since its collision with Asia?

A: There is evidence to suggest a continued northward motion of the India plate for a distance of about 2,000 km at a rate of about 5 cm a year.

Q: In what way is this continued motion responsible for the present earthquake activity?

A: The squeezing of Asia north of the Himalayas is producing large-scale faulting, with displacements of tens of meters along the major faults. It is clear that the region is not acting as a rigid plate.

Q: What are some of the greatest faults in the region?

A: Perhaps the longest strike-slip fault in the world is the Altyn Tagh fault north of Tibet. It can be traced on satellite photographs for more than 2,500 km, if the Kansu fault, which merges at its eastern end, is included. Another great strike-slip fault, the Bolnai fault in northern Mongolia, ruptured for 370 km in a great earthquake in 1905. Many other major strike-slip faults have now been mapped in the area.

Q: Why does the oceanic rather than the continental lithosphere subduct?

A: The continental crust, which makes up the upper part of the lithosphere, is about 35 km thick and is 20 percent less dense than the mantle. Its buoyancy probably keeps it from being subducted. In contrast, the oceanic crust is only about 5 km thick and is 15 percent less dense than the mantle rocks. Its lower buoyancy favors subduction.

Q: What happens when two continental plates collide?

A: From the evidence of the Tibetan collision and the Alpine collision in the Mediterranean, permanent attachments of continental crusts occur through complicated crustal deformations and mountain building. The term *to suture* is used to describe this complicated attachment process.

Q: How do rocks called *ophiolites* define a suture between two continents?

A: The suite of rocks called *ophiolites* are found in the zone of suture between two continents. Ophiolites are a sedimentary and igneous sequence that includes cherts (characteristic of deep ocean deposition), pillow basalts (typical of lava flows on the ocean floor), and ultramatic rocks (typical of the upper mantle). This mixture of rocks indicates that zones of suture were previously ocean basins.

Q: What are the main types of fault displacement?

A: When the rock on the side of the fault hanging over the fracture slips downward, the faulting is called *normal*. When the hanging wall of the fault moves upward, the fault is called a *reverse fault* or *thrust fault*. Normal and reverse faults produce vertical displacements and are called *dip-slip* faults. In contrast, when only horizontal slips occur along the strike of the fault, the fault is called *strike-slip*. Motion on a fault may be a combination of dip-slip and strike-slip.

Q: What is the major tectonic problem of the India-Asia collision?

A: We must explain how a piece of crust 2,000 km long and the width of India that was once located in the region of India's northwest advanced into Asia during the last 40 million years.

Q: What hypotheses have been put forward to explain these large displacements?

A: There are three main hypotheses. The first is that the north margin of India has thrust under the Tibetan plateau, producing an uplift of the Himalayas. Focal mechanism solutions of earthquakes along the southern edge of the Himalayas do indicate thrust mechanisms. The second hypothesis is that the Tibetan plateau has been forced upward through crustal shortening, in which the contraction has been horizontal and the expansion vertical, like the motion of an accordion. Calculations indicate that, if the crustal thickness is doubled in Tibet and Mongolia due to the north-south pressure, the squeezing would account for about 1,000 km of India's total travel toward Eurasia since the beginning of the collision. The third hypothesis is that India acted like a blunt punch that indented Asia, not only by squeezing mountains up in front of it but by pushing material out to the sides along the left lateral strike-slip faults. In particular, most of the 2,000 km crustal shortening is attributed to lateral displacement of China toward the east over the last 40 million years.

Q: How have fault plane solutions of earthquake mechanisms in Asian earthquakes helped to distinguish between the hypotheses?

A: The first P wave to arrive at a seismographic station from the focus of an earthquake arrives as a push (compression) or pull (dilatation) of the ground, depending on the actual motion of the fault plane at the focus. Thus, patterns of pushes and pulls at seismographic stations around the world can be used to calculate whether the faulting is normal, thrust, or strike-slip. Fault plane solutions of Himalayan earthquakes are usually of thrust type. This is also true for the Tien Shan region to the west of the area. In contrast, fault plane solutions of a great many earthquakes in Mongolia indicate strike-slip faulting consistent with a northeast-southwest orientation of the maximum compressive stress.

Q: Are there any notable exceptions to the general compressive stress pattern of central Asia?

A: Although generalizations on overall compression can be made in most parts of the region, faulting and fault plane mechanisms are somewhat complex. There are as well two notable exceptions to the horizontal compression model. These are the Baikal and Shansi rift systems. Both of these rift valleys, associated with normal earthquakes, are created by extension of the crust under tensional forces.

Q: How can a crucial test be made of the hypothesis that the India-Asian convergence is associated with large lateral displacement of China?

A: If it can be shown by field geological studies that horizontal displacements along the great strike-slip faults of Asia have been limited to only a few tens of kilometers rather than hundreds of kilometers during the last 40 million years, the hypothesis would be untenable.

Q: Can plate tectonics be applied directly to the explanation of the evolution of the geological structure of Asia?

A: The building of great mountains, plateaus, and rift valleys in this area does not follow from a direct application of plate tectonics to the continental crust and lithosphere. Many minor or nonrigid plates would have to be introduced to provide a detailed mechanical explanation of the complex deformation of the area.

Q: How do seismologists use the travel times of seismic waves, such as P and S waves, to determine the structure of the earth's interior?

A: The time that it takes a P or S wave to travel from its generation at a fault to a seismographic station at a remote part of the earth depends on its velocity and whether it has encountered any reflecting surfaces that have changed its direction. By comparing the arrival times of seismic waves of various types at many stations on the earth's surface, seismologists can determine the speed at any depth in the earth and the location of reflecting discontinuities in structure.

Q: How can determination of the velocities of seismic waves through the earth give information on its physical properties?

A: The most commonly used method is to note that the velocity of P waves is equal to a function of the incompressibility, elastic rigidity, and density of the materials the waves pass through. The velocity of seismic S waves is a function only of the elastic rigidity and density of the rocks. Thus, the speed of *P* and *S* waves at a certain depth gives information on the ratio of any two of these parameters. If *S* waves do not propagate, the material is a liquid (zero rigidity).

Q: What is the seismological evidence for a distinct core of the earth?

A: The most direct evidence is the observation of earthquake waves called *PcP* waves recorded at stations not too distant from the seismic source. The time of travel of these waves indicates that they have propagated almost vertically downward in the earth and have been reflected back to the recording station from a rather sharp discontinuity at a depth of about 2,900 km. Other strong evidence concerns the behavior of seismic waves that are recorded at distances over 100° (the angular distance subtended at the center of the earth) from the wave source. At greater distances, the amplitude of short-period *P* waves begins to decrease, and these stations are clearly in a shadow of some obstacle about halfway to the center of the earth that has reflected and diffracted the waves away from a direct path. *S* waves reflected from the core boundary, called *ScS*, are also observed.

Q: What is the evidence that the outer core is liquid?

A: The most direct evidence of solid rocks in the earth's interior is the observation of seismic *S* waves that have passed through them, because these waves cannot propagate in a liquid. *S* waves are seen at all distances through the mantle of the earth, from angular distances of 0° to distances of at least 100°. By contrast, seismologists have never been able to detect *S* waves that have passed through material deeper than the mantle-core boundary. This result lends much support to the inference of a liquid core. However, if hot rocks at those depths damped out *S* waves very rapidly, they would be difficult to detect. More direct evidence for a liquid outer core is found from the tidal response of the earth as a whole and from the free oscillations of the earth.

Q: What is the evidence for the inner core of the earth?

A: Just as with the *PcP* reflections from the outer core, seismograms show that certain *P* waves must be reflected from some object deep within the core with a diameter about that of the moon. These reflections, called *PKiKP*, may be tracked back from 140° to within 10° of the earthquake epicenter; the character of the recorded waves indicates that the outer boundary to the inner core is quite sharp.

Q: What is the evidence that the inner core is solid?

A: Converted *P* waves that have passed as *S* waves through the inner core have been reported on seismograms, but this identification has not been generally confirmed. The evidence for a solid inner core is therefore based on two main arguments. First, *P* waves that have passed through the inner core have a greater velocity within it than within the liquid outer core. This greater velocity can best be accounted for by assuming that the material of the inner core is rigid. The second argument comes from matching observed periods of free oscillations of the earth with theoretical periods calculated from assumed models of the earth. These matches are closest when a liquid outer core and a solid inner core are incorporated in the model.

Q: Is there evidence for fine structure near the top of the earth's mantle?

A: Evidence for a small but sharp change of properties at depths of about 400 km and 650 km has accumulated over the years. Some of the new evidence comes from waves that have reflected from the underside of these boundaries, such as the waves *P'dP'*. These have traveled from one side of the earth through the earth's core and have been reflected back at these structural changes before reaching the surface at the other side of the earth. Finally, they have been recorded at stations as they travel upward again from the interior.

Q: What is the evidence for a transition shell (D'') around the core at the base of the mantle?

A: The direct P waves that travel to the surface at distances of about 100° show properties that depend on their wavelengths. At greater distances, short-period waves have a time of arrival curve that is a straight line. The slope of this curve indicates velocities just above the mantle-core boundary that are less than the velocities obtained at a distance of 100 km above the mantle-core boundary. Similar evidence can also be found from S waves that have been diffracted around the core of the earth. These S waves have velocities about 2 percent less than those 150 km above the mantle-core boundary.

Q: What physical interpretation can be given to the mantle-core (D'') transition shell?

A: There are two main explanations, both of which may be partly true. One is that the decrease in seismic velocities is a result of increased density in this shell, perhaps due to the diffusion outward from the earth's liquid core of a small amount of liquid iron. The second interpretation is that the temperatures rise rapidly in this zone between the lower temperatures of the mantle and the higher temperatures of the outer core.

Q: What are the multiple reflections $PK7KP$, and what do they indicate about the nature of the mantle-core boundary and the attenuation properties of the material of the outer core?

A: Waves denoted $PK7KP$ refer to P waves that have entered the outer core and have been reflected six times on the inside of the mantle-core boundary before finally emerging again through the mantle and being detected at the earth's surface. Many phases of this type, such as $PKKP$ up to $PK13KP$ (with twelve reflections), have been observed in recent years. The seismograms show that the pulses of this type of wave are not changed significantly in shape, even after many reflections, and that their travel times fit closely the predicted times based on single reflections. Both properties indicate that they are being reflected at a very sharp boundary as compared with their wavelength. The slow diminution in amplitude of these waves, even after many reflections, indicates that the attenuation properties of the liquid outer core are very small relative to those of the earth's mantle.

Q: How can measurements of the amplitudes of seismic waves at the earth's surface give information on the density of the inner core?

A: The amount of energy that is reflected at a sharp boundary, such as that at the inner core when a P wave is incident upon it, is a function of the ratio of the elastic properties and density on each side of the boundary. If it is assumed that the elastic properties and density of the mantle and outer core are already known, then the amplitude of the reflections $PKiKP$ at the earth's surface gives direct information on the change in elastic properties and densities at the inner core boundary. Because the outer core is a fluid and the vertically incident P waves do not produce converted S waves in the inner core, the remaining unknown term is the density of the inner core.

Q: In what ways have the efforts to obtain a comprehensive nuclear weapons test ban treaty improved seismology?

A: To monitor such a treaty adequately, an observational base must be secured. Thus, following recommendations made in 1960, an upgraded worldwide network of standardized seismographs was installed in many countries. Even countries that did not participate in this observational program took steps to improve recording equipment, modernize seismographs, and keep accurate time at seismographic stations. Also, special equipment was developed and placed around the world in the form of seismic arrays, or linked groups of seismographs. These seismic arrays allow the filtering out of background vibrations that obscure the small seismic signals from distant sources.

Q: What is meant by deep root zones under the older parts of continents?
A: The concept is of structural regions that extend deep below the lithosphere and into the asthenosphere. These regions are postulated to be particularly thick, reaching perhaps to 400 km under the continental shields, and it is argued by T. H. Jordan that they may be attached to the old continents, like the roots of teeth, so that the deep zones and the continents drift together.

Q: What is the evidence from earthquakes for deep continental roots?
A: The earliest evidence came from the observed contrast in the wave dispersion of long-period seismic surface waves between paths across oceans and continental shields. This evidence suggested differences in structure below continents and oceans down to depths of 200 km or so. Recently, measurements from S body waves that pass almost vertically through the upper mantle have provided more resolution. Travel times of multiply reflected ScS waves from the core boundary are 4 seconds greater under ocean basins than expected under old continents. This large difference can be explained if velocities in the rocks under continents are higher than under oceans to depths of 300 to 400 km.

5

The Collision between India and Eurasia

by Peter Molnar and Paul Tapponnier
April 1977

*For the past 40 million years the Indian subcontinent
has been pushing northward against the Eurasian land
mass, giving rise to the severest earthquakes and the
most diverse land forms known*

The region of the earth that exhibits the greatest diversity of geology, topography and climate, together with a strong susceptibility to major earthquakes, is the part of Eurasia that lies east of the Ural Mountains and north of the Ganges, embracing northern India, Pakistan, Afghanistan, the Tibetan plateau, Mongolia, most of China and a large part of the eastern U.S.S.R. This profoundly disarranged region is roughly equal in area to all of North America from the Rio Grande to 60 degrees north latitude, the parallel that crosses the northern tip of Labrador on the east and Cook Inlet in Alaska on the west. The world's highest mountains, the Himalayas, rise abruptly from the flat, densely populated Ganges Plain in northern India and shield the Tibetan plateau from the seasonally shifting monsoon winds of southern Asia. Tibet supports a population of less than 1.5 million in an area somewhat larger than Texas, Oklahoma and New Mexico combined. Tibet's average elevation is 5,000 meters, higher than any point in the 48 contiguous states of the U.S. In contrast, eastern China has abundant rainfall and supports a population of close to a billion. (As long ago as 1556 a single earthquake near Sian, then the capital of China, is known to have killed 830,000 people.) The broad, high mountainous zone that includes the Himalayas and the Tibetan plateau presents such a barrier to travel that throughout history the populations of India and China have had remarkably little contact. North of Tibet the Gobi Desert presents another formidable barrier to migration and communication. The Tarim Basin, part of the Gobi Desert, is one of the driest and most inhospitable regions on the earth. Mountains as high as 5,000 meters surround it on three sides. Nearly constant winds blow across the basin with such force that they pile up sand dunes as much as 150 kilometers long with wavelengths of three to five kilometers, clearly visible in

satellite pictures. North of the Gobi Desert, Lake Baikal, at 1,800 meters the deepest lake in the world, fills the Baikal rift zone, a huge crack in the earth's crust similar to the one in East Africa.

Although the geology of Asia seems to present a chaotic jumble of land forms, much of the deformation of the surface, when it is viewed as a whole with the help of satellite photographs, seems to fall into a simple, coherent pattern attributable to a single cause: a geological collision between the Indian subcontinent and the rest of Eurasia. This collision is still in progress. As an example of its current effects, we believe the great earthquake that devastated the industrial city of Tang Shan last summer was the result of forces originating in the collision area 2,500 kilometers to the southwest.

The possibility of quantitatively analyzing the collision between India and Eurasia has developed only in the past 10 years. The hypothesis of sea-floor spreading was first advanced in the mid-1960's to explain the mid-ocean ridge, the great mountain range on the ocean floor that runs for 40,000 kilometers across the world's ocean basins. At a rift in the crest of the ridge molten rock wells up from the mantle below and fills the gap that is left as the ocean floor on each side of the rift moves outward. It was also recognized that the ocean floor is magnetized in stripes of opposite polarity, depending on the magnetic polarity of the earth itself at the time the molten rock crystallized. (For reasons still not known the earth's magnetic polarity reverses at intervals of several hundred thousand years.) The study of magnetic reversals in the floors of the Indian and Atlantic oceans showed that India had traveled 5,000 kilometers with respect to Eurasia before the two collided. Since the beginning of the collision India has continued to move northward another 2,000 kilometers with respect to Eurasia. How can one account for the vast area of land that was displaced by the

collision? We shall describe our hypothesis to account for the displacement and present the evidence supporting it.

The recognition of sea-floor spreading, which confirmed earlier conjectures of continental drift, quickly led to the much broader concept of plate tectonics, a concept that has inspired the renaissance in the earth sciences. Plate tectonics provides a physically simple mechanism for large-scale horizontal motions of separate portions of the earth's crust and makes it possible to be quantitatively precise in describing the kinematics of continental drift. One of the central concepts of plate tectonics is that a small number of large plates of the lithosphere, the high-strength outer shell of the earth, move rigidly with respect to one another at rates of one centimeter to 20 centimeters per year over the hotter, low-strength asthenosphere under them. Some 100 kilometers thick, this outer shell consists of the earth's crust and the upper part of the mantle. Thus in plate tectonics the earth's crust, which is both lighter than and chemically different from the mantle under it, is visualized as being carried passively as part of a lithospheric plate.

It is ironic that Alfred Wegener, who was probably the most important early proponent of continental drift, based much of his argument on the difference between the continental crust and the oceanic crust. The continental crust, which stands high above the oceanic crust, has deep roots going down about 35 kilometers, whereas under the oceans the crust is only about six kilometers thick. Wegener viewed the continents as sturdy ships sailing majestically through the much weaker crust and mantle under the oceans. In actual fact, as is recognized in plate tectonics, the oceanic crust and upper mantle—the lithosphere—are extremely strong. They seem to be deformed only at the boundaries of plates, so that the relative motion of the plates can be described as the motion of rigid bodies.

The continents usually act as rigid structures when they lie entirely within one plate. When a boundary between two plates passes through a continent, however, it is generally more diffuse than the boundary between two oceanic plates or the boundary between an oceanic plate and a continental one, and it is accordingly much more difficult to define. Such diffuseness is apparent throughout the Mediterranean area, and it is particularly evident in Asia, where the India and Eurasia plates are converging.

Before proceeding further we should like to emphasize the important role played in plate tectonics by the difference between continents and oceans. Oceanic crust is formed as a part of the lithosphere at spreading ocean ridges. The oceanic lithosphere cools, shrinks and gets denser as it moves away from

PORTION OF ALTYN TAGH FAULT, photographed from an altitude of 950 kilometers by the Earth Resources Technology Satellite (ERTS) in February, 1973, is some 1,200 kilometers north of Mount Everest in the Sinkiang region of China. The fault, which slants to the upper right corner from the lower left, may be the greatest active continental strike-slip fault in the world. The term strike-slip means that the two sides of the fault are sliding past each other in opposite directions. The southern block of the Altyn Tagh fault is evidently being driven to the right, or eastward, with respect to the northern block as a consequence of the collision of India with the southern margin of Eurasia. The collision began some 40 million years ago and is continuing. About 180 kilometers of fault appears here. Entire fault can be traced for more than 2,500 kilometers if one includes Kansu fault, with which it merges at its eastern end (*see map on pages 8 and 9*). This ERTS photograph and those on pages 10 and 11 are reproduced with south at the top, so that topographical features are illuminated from the top. Otherwise shadows would point upward and the relief would tend to reverse so that the valleys would appear to be ridges.

the spreading center. An amount of crust equivalent to the amount added at a spreading ridge must return continuously to the mantle. This happens at subduction zones, where the oceanic lithosphere plunges into the asthenosphere, pulled down because the descending lithosphere is cooler and therefore slightly denser than the asthenosphere surrounding it. Although the oceanic crust is about 15 percent less dense than the mantle, it is sufficiently thin to be carried down as part of the lithosphere. This appears to happen in many places, typically forming deep trenches adjacent to island arcs such as those rimming the western Pacific.

Consider, however, what happens when a continent riding on the downgoing slab of lithosphere tries to plunge into the asthenosphere, traveling behind the consumed oceanic lithosphere. The continental crust, some 35 kilometers thick and almost 20 percent less dense than the mantle, is both too thick and too light to be carried down into the asthenosphere. Thus the buoyancy of the continental crust keeps it from being subducted and probably explains the fact that whereas the oceanic crust recycles itself about every 200 million years, large areas of the continents have existed for billions of years.

When two continents collide, they suture themselves together to form a larger continent. Either the relative motion of the two plates on which the continents ride will change or a new boundary will form between the plates at a different place. Geologic history provides several examples. Some 250 million years ago Europe and Siberia were sutured at the Ural Mountains, and at about the same time North America and Africa were sutured at the southern Appalachians, before the present-day Atlantic Ocean formed. When one looks at a geological map of Asia, several ancient sutures are apparent. A few years ago Peter N. Kropotkin, the first well-known Russian geologist to accept continental drift (and the great-nephew of the famous anarchist of the same name), described Eurasia as a "composite continent" to which fragments have been successively sutured by collision over the past 800 million years. Most of the sutures in Eurasia appear to be older than 200 million years; only the suture between the Indian subcontinent and the rest of Eurasia appears to be younger. Here we shall consider only this most recent collision and its consequences.

Although we argue that typical plate tectonics does not seem to apply to the portion of Asia that lies between India and Siberia and is tectonically active today, plate tectonics can be used to place constraints on the history of the relative

motion of the India and Eurasia plates. We can exploit the fact that if we know the history of the relative motion of plate a and any two other plates, b and c, we can calculate the history of the relative motion of b and c. From geological studies in the North Atlantic by Jean Francheteau of the Centre Océanologique de Bretagne in France and by Walter C. Pitman III and Manik Talwani of the Lamont-Doherty Geological Observatory of Columbia University we know how both Eurasia and Africa moved with respect to North America, and thus we can calculate how they moved with respect to each other. Similarly, from the work of Robert L. Fisher of the Scripps Institution of Oceanography, D. P. McKenzie of the University of Cambridge and John G. Sclater of the Massachusetts Institute of Technology we know how India and Africa moved with respect to each other and therefore we can calculate where India was with respect to Eurasia at different times in the past.

We do not know where the northern margin of the original Indian continent lies with respect to India today, because that margin has been much deformed in the creation of the Himalayas. For the purposes of calculating how much the India and Eurasia plates moved with respect to each other in different intervals, however, our ignorance on this point does not matter. Moreover, although we do not know how far north of India the old margin lies, we can determine from the geology of Asia the position of the boundary between rocks of the present Indian subcontinent and those that were part of Eurasia long before the collision.

The primary evidence used to delineate a suture between two continents is the existence of the sequence of rocks known as an ophiolite suite. Ophiolites have three distinguishing characteristics. They show a particular sedimentary sequence incorporating bedded cherts, which are characteristic of deep-ocean sedimentary deposition. They contain remnants of pillow basalts: igneous rocks of lumpy form that are typical of basaltic lava extruded under water, as at spreading ocean ridges. And they incorporate dense, dark rocks low in silicon dioxide known as ultramafics, which are thought to be typical of the mantle. Ophiolites are interpreted as being a slice of the oceanic crust and upper mantle; accordingly their presence implies the former existence of an ocean basin.

A belt of ophiolites follows the Indus and Tsangpo valleys in southern Tibet north of the Himalayas; it appears to mark the boundary between the sutured continents. Somewhat farther north there are volcanic rocks typical of those found at subduction zones, as in the Andes of South America. Just as oceanic

COLLISION between India and Eurasia appears to have compressed and distorted the earth's crust from the Himalayas to Siberia and from Afghanistan to the coast of China. Immediately north of the Himalayas, Tibet has been raised to an average elevation of 5,000 meters. The Tibetan plateau is drained

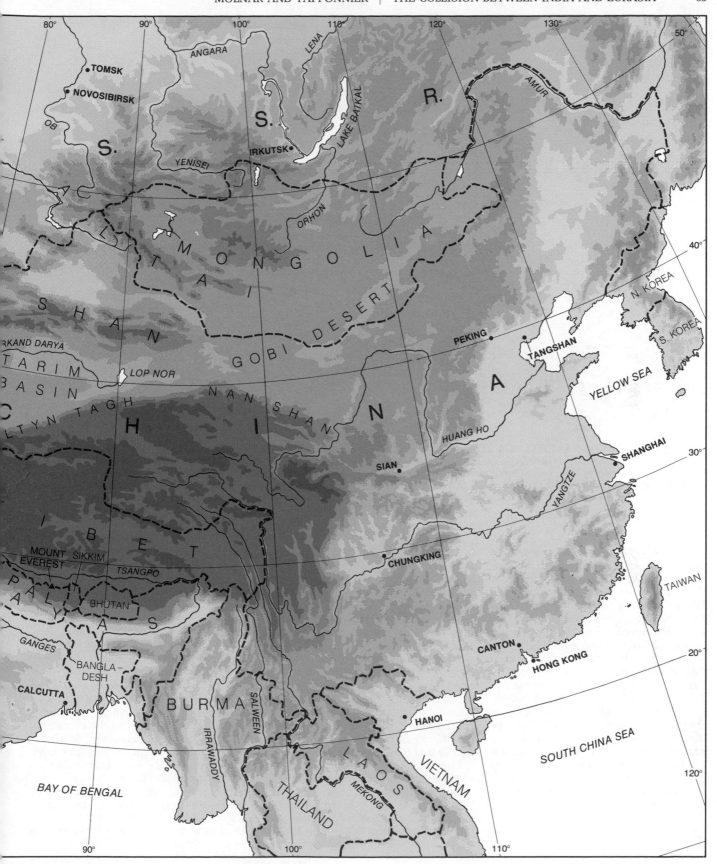

to the east by a series of great rivers that arise in nearly parallel channels but eventually fan out into rich deltas along an arc extending from the Bay of Bengal to the Yellow Sea. To make the geography and distances more familiar one can imagine a map of North America traced on top of this one so that the Rio Grande is roughly aligned with the Ganges, which would place Brownsville, Tex., close to Calcutta. (Actually Calcutta is about four degrees farther south than Brownsville.) Los Angeles would then coincide approximately with Kandahar in Afghanistan, Miami would be near Hanoi, and Portland, Me., would be close to Peking. The Himalayas would sweep in a great arc from Nevada-Utah border to Mississippi, crossing Arizona, New Mexico, Texas and Louisiana. Tibetan plateau would include Colorado, Nebraska, Kansas, Oklahoma, Iowa and Missouri. Far to the north Lake Baikal would cut across northernmost tip of Quebec.

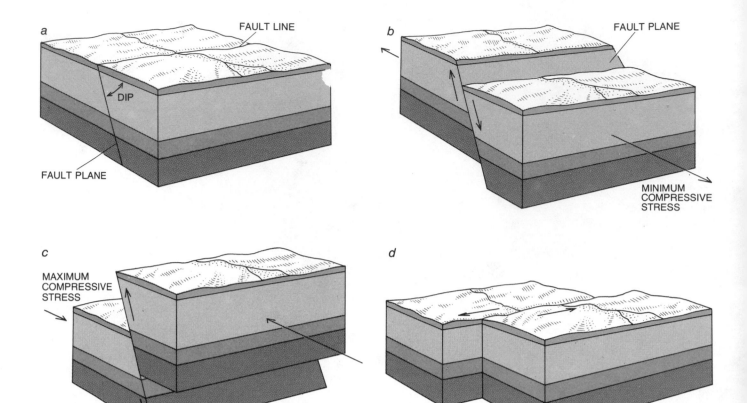

DURING AN EARTHQUAKE one block of the earth's crust slips with respect to an adjacent block along a fault plane. In a normal fault (*b*) the blocks act as if they were being pulled apart. The overlying block slides down the dip of the fault plane. In a thrust fault (*c*) the overlying block is forced up the dip of the fault plane because the maximum compressive stress is horizontal and perpendicular. In a strike-slip fault (*d*) the two blocks slide past each other. Lateral slippage can be combined with normal or thrust faulting. From an analysis of the seismic waves generated by an earthquake, called a fault-plane solution, one can tell what kind of fault motion has occurred.

crust now plunges to the east under the Andes, presumably the old ocean floor between India and Eurasia plunged to the north under Tibet. In contrast the Himalayas, south of the suture, consist of slices of the old northern portion of India that have been stacked one on top of another to form the mountains. Patrick LeFort of the Centre de Recherches Pétrographiques et Géochimiques in France and Maurice Mattauer of the University of Montpellier have shown that there is a progression from north to south in the piling up of such slices, so that the oldest thrust is to the north. The next step will probably be for a new fault to form farther south on the Ganges Plain and for material to the north of it to slide up and over the plain to the south.

Although the geology of the Himalayas and Tibet is not well enough known to enable us to reconstruct the position of the northern margin of the original Indian continent, it does place important constraints on when India and Eurasia could have collided. Four observations imply a date of between 40 and 60 million years ago. First, from geological investigation of the ophiolites in the area, Augusto Gansser of the Swiss Federal Institute of Technology

found "exotic blocks" of the late Cretaceous (about 70 million years ago) within the suite. Thus the ophiolites must have been part of an ocean floor until after that time.

Second, Gansser describes a sequence of sedimentary rocks on the northern edge of the Himalayas that is typical of sequences found on continental shelves and slopes and that begins in the Cambrian (about 500 million years ago) and continues until the early Eocene (about 55 million years ago). Hence there appears to have been an ocean between the converging continents until approximately that time.

Third, there is no known fossil record of mammals in India before about 50 million years ago. Ashok Sahni and Vimal Kumar of Lucknow University in India report that the oldest mammals in India date from the middle Eocene (about 45 million years ago). The first Indian mammals are similar to those found in Mongolia. Thus although mammals had evolved on other continents, it appears that they did not evolve independently in India, indicating that the continent remained isolated until about 45 million years ago. The collision enabled a horde of Mongolian mammals to sweep into India. Fourth, Gansser infers that major mountain

building in the Himalayas began in the Oligocene (about 35 million years ago).

These observations do not enable us to determine precisely when India and Eurasia collided, but we think the event probably occurred 45 million years ago, give or take a few million years. It is highly improbable, however, that the old margins of India and Eurasia met flush along their full length. It is more likely that they first made contact when peninsulas met and that with the passage of time the zone of contact grew until the ocean basin between the continents was swallowed up. Thus we consider it likely that the initial contact may have been a few million years earlier than intimate contact.

From the history of sea-floor spreading in the Indian Ocean one can calculate the relative positions of India and Eurasia over some tens of millions of years. One can see immediately that between 70 and 40 million years ago India moved about twice as far with respect to Eurasia as it has since then. A plot of the distance of the northeast and northwest corners of India at different times in the past from their present positions shows that the rate of convergence between them changed by a factor of two about 40 million years ago. Given the uncertainties in the data the change in rate could have come a few million years later or as much as 10 million years earlier. In any event we interpret the change in the rate at which India approached its present position as an indication that the first stages of collision came at about the same time as the change in rate, and that the buoyant continental crust of India, instead of being subducted, put a brake on the northward motion of the India plate.

Although the reconstructions provide support for the view that continental crust cannot be subducted, they leave us with what may be a more difficult problem. If we conclude that India and Eurasia collided 40 million years ago, we must also conclude that since then India has traveled northward about 2,000 kilometers with respect to Eurasia. If the continents collided earlier, the distance covered is even greater. Bearing in mind that continental crust cannot be subducted, we are faced with the problem of accounting for the displacement of a piece of crust that has an area the width of India and is 2,000 kilometers long.

The continuing northward motion of India at a rate of about five centimeters per year is probably responsible for the widespread tectonic activity in Asia. For example, seismic activity is detected over an area extending some 3,000 kilometers north and east of the Himalayas. Among the 22 greatest earthquakes listed by Beno Gutenberg and Charles F. Richter of the California Institute of Technology for the period 1897–1955, seven occurred in central and eastern Asia, four of them north of the Himalayas. The deformation of the surface of the earth that accompanied some of these earthquakes was huge. The 1957 Gobi-Altai earthquake in Mongolia, which came after the Gutenberg-Richter study and would probably have been too small to have been included, caused displacements of as much as 10 meters along the main fault associated with the earthquake. Such large displacements along faults seem to have been characteristic of several of the great earthquakes in Asia. In any case it is clear that the region as a whole does not act as a rigid plate.

Asia is known for high mountains not only in the Himalayas and Tibet but also farther north and east, where in the Tien Shan and Nan Shan ranges there are peaks of up to 6,000 meters. Ordinarily high mountains are rapidly worn down by erosion, so that their existence implies large crustal movements in recent geologic times. Studies conducted by Russian geologists (V. N. Krestnikov, A. V. Goryachev, S. A. Zakharov and others) show that the area of the Tien Shan range was nearly flat from 200 million years ago to 30 or 40 million years ago, and that it has been elevated since then. Although other reports we have seen are less definitive, they do suggest that the relief in Mongolia and China

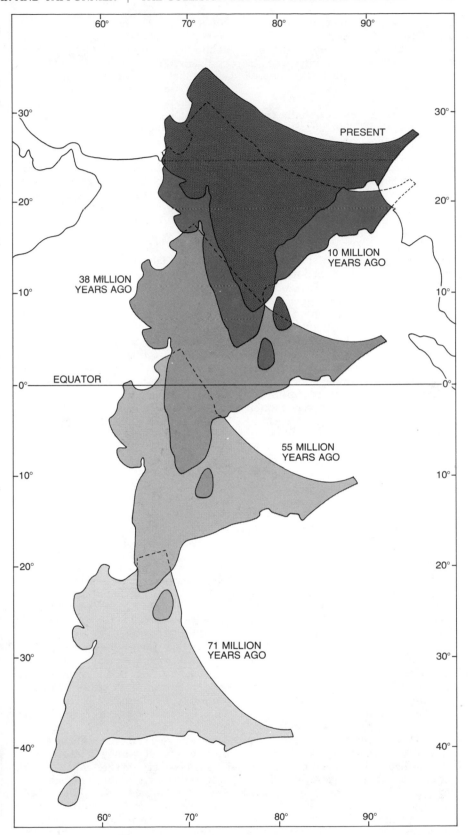

INDIA'S NORTHWARD DRIFT has been reconstructed from magnetic reversals in the floors of the Indian and Atlantic oceans. As molten rock welled up into the rift in the ocean floor and hardened it became magnetized according to the prevailing polarity of the earth's magnetic field. At infrequent and irregular intervals the earth's polarity changes, leaving a record that can be dated. This "time lapse" reconstruction shows that India traveled some 5,000 kilometers northward with respect to Eurasia in the 20 to 30 million years before its collision with Eurasia. Over the past 70 million years the northeastern tip has actually traveled some 7,000 kilometers. Velocity for continent as a whole was about 10 centimeters per year for the first 30 million years and about five centimeters per year for the next 40 million. In this reconstruction it is arbitrarily assumed that the boundary of Eurasia is fixed in its present location.

today is comparably new. Hence the deformation of a large part of the crust of Asia appears to have begun after the collision with India. We infer that the penetration of India into Eurasia caused the deformation.

In trying to explain how India could have traveled 2,000 kilometers after it began to collide with Eurasia we obviously have greater freedom in accounting for the material displaced if we imagine its being absorbed over several million square kilometers rather than in a narrow zone near the suture. The formation of the Himalayas can account for only a fraction of the material displaced, so that it remains necessary to explain where the rest of it has gone. One method that has proved to be particularly useful for deciding how material is displaced in oceanic regions at present is the determination of "fault-plane solutions" of earthquakes. Such solutions reveal the direction of relative movement along the fault. From the study of the waves radiated by an earthquake one can determine both the type of faulting that took place during the earthquake and the direction of motion of one side of the fault with respect to the other.

Geologists classify faults in three main categories. A normal fault results when stresses directed horizontally cause one side of the fault to sink with respect to the other. A thrust fault is produced when compressive stresses drive one side of a fault over the other side. In a strike-slip fault the two sides of a fault slide horizontally past each other. If the opposite side of the fault moves to the left, as viewed from either side, the movement is termed left-lateral displacement. Movement in the opposite sense is right-lateral displacement. Working with Thomas J. Fitch, who is now at the Lincoln Laboratory of M.I.T., and Francis T. Wu of the State University of New York at Binghamton, we have compiled fault-plane solutions for some 75 earthquakes in Asia.

Studies of earthquakes in the Himalayas corroborate geological observations and show that India is continuing to thrust under the Himalayas in a northerly to northeasterly direction. Over the past 50 years several investigators have suggested that the northern margin of India was also thrust under the Tibetan plateau and that Tibet is at a higher altitude because of it. We and many other geologists now find this very unlikely. We see no evidence of such underthrusting going on now, and mechanically such a phenomenon seems highly contrived.

More recently John F. Dewey and Kevin Burke of the State University of New York at Albany have suggested that the Tibetan plateau is a result of a shortening of the crust in the entire Ti-

betan-Himalayan region, with the Tibetan crust behaving like an accordion, contracting horizontally and expanding vertically. Although we are not convinced that this concept is inapplicable to Tibet, we see no persuasive evidence for accepting it. For one thing, studies of earthquakes show that Tibet is not shortening but is stretching in an east-west direction. Moreover, from a comparison of satellite photographs of Tibet with those of other mountainous areas in Asia where shortening is currently taking place, the surface deformation of the Tibetan plateau appears to be much less, and if folding of the crust has occurred, it is less recent. Nevertheless, even if the thickness of the crust has doubled in Tibet as a result of pressure from the south, only 600 or 700 of the 2,000 kilometers can be explained in this way.

In the Tien Shan area the pattern of deformation is again dominated by thrust faulting, with shortening in a northerly direction. The Russian seismologist V. I. Ulomov estimates that the amount of shortening in the western part of the area is about 300 kilometers, a figure arrived at by imagining what would happen if the thickened crust there was flattened out (in his words "with a rolling pin") to normal thickness. For most of the rest of Asia fault-plane solutions of earthquakes indicate the overall predominance of strike-slip faulting. Thrust faulting is characteristic in only limited areas; normal faulting is fairly uncommon. For example, in Mongolia most earthquakes are associated with strike-slip faulting. In any case earthquakes in Mongolia exhibit a fairly consistent northeast-southwest orientation of the maximum compressive stress. Although the pattern east of Tibet is more complex, it is similar to the pattern in Mongolia.

Thus the faulting associated with earthquakes indicates that much of Asia is being squeezed in a direction lying between north-south and northeast-southwest, a pattern that is compatible with India's northward motion. The squeezing causes the crust in parts of Asia to shorten and thicken. Here again, however, even if one adds a possible 700 kilometers of crustal shortening in Tibet to 300 kilometers in the Tien Shan area, one can still account for only 1,000 kilometers of shortening, or half of India's total travel toward Eurasia since the beginning of the collision. Another explanation must be sought for the remaining 1,000 kilometers of displacement even if we are wrong in doubting the "accordion" shortening in Tibet.

The satellite pictures of Asia play a central role in leading us to an alternative explanation for the 2,000 kilometers of convergence between India and Eurasia. Perhaps the most striking fea-

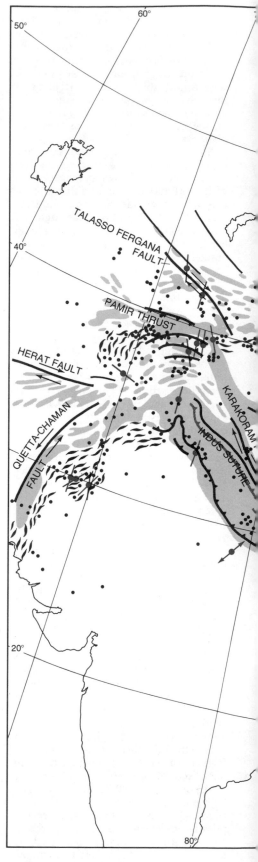

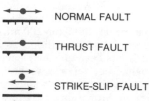

NORMAL FAULT

THRUST FAULT

STRIKE-SLIP FAULT

 FOLDS

COMPRESSION ZONES

EXTENSION ZONES

· SMALL EARTHQUAKES

● LARGE EARTHQUAKES

PRINCIPAL TECTONIC FEATURES that are thought to be associated with continuing northward push of the India plate against the Eurasia plate have been plotted by the authors, partly on the basis of the analysis of ERTS photographs and partly on the basis of studies of major earthquakes (*colored dots*), which reveal how the crust has moved along faults. The straight lines without arrowheads through dots indicate thrust faults. The double-headed arrows indicate normal faults. The pairs of antiparallel arrows indicate movement along strike-slip faults. The areas in color appear to be zones of recent uplift resulting from crustal shortening. The overall impression is that the large Eurasian land mass that lies to the west of 70 degrees east longitude has remained more or less undeformed as China has been pushed to east.

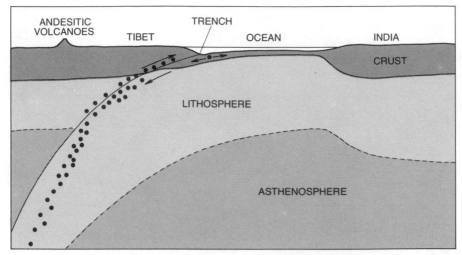

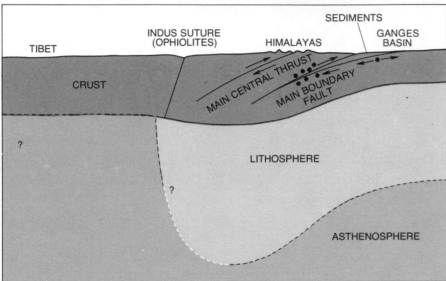

CROSS SECTION OF THE COLLISION between India and Eurasia plates is shown schemat- ically. The upper diagram shows a cross section through the lithosphere and the asthenosphere about two million years before actual contact, when the land masses were still separated by about 200 kilometers of ocean. At that time the lithospheric plate carrying India was plunging under the Eurasia plate as today the Pacific plate is plunging under the South America plate, creating the Andes along the west coast of South America. The black dots show how earth- quakes tend to cluster along the boundary between plates and within the descending plate. The lower diagram depicts the situation today. The suture line, the Indus suture, is marked by the presence of ophiolites: sequences of rocks containing ocean sediments and showing other char- acteristics of having been formed in a suboceanic environment. Earthquakes are more dif- fusely distributed and shallower than they were before the collision. The Himalayas are slices of old Indian crust that have overthrust rest of India to the south, creating new faults that mi- grate southward. Active faulting seems to occur on main boundary fault. Crust under Tibet ap- pears to be unusually hot; lithosphere there may be so thin that bottom may lie within crust.

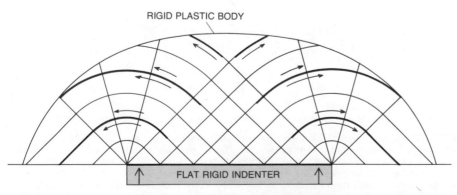

GEOMETRY OF SLIP LINES observed when an indenting tool made of a hard material such as steel is pressed into a softer material such as brass bears a striking resemblance to the distri- bution and directional sense of strike-slip faults in Asia. The slip lines are probably less sym- metrical in Asia than they are in idealized case because of asymmetry of boundary conditions.

tures seen in aerial photographs of any kind are strike-slip faults. They usually appear as sharply defined linear features in the topography. In Asia the strike-slip faults are among the clearest in the world. Although some of these faults were known from displacements during earthquakes or from geological studies, it appears that until recently some were not known, at least to most geologists.

The Talasso Fergana fault in the Tien Shan area is well known to Russian geol- ogists and also stands out clearly in sat- ellite pictures. The sharp bends in the ridges near the fault trace distinctly show that the sense of motion for the fault is right lateral. The Bolnai, or Khangai, fault in northern Mongolia ruptured in 1905 during one of the greatest earthquakes of the century. Al- though the Bolnai fault is little known outside Mongolia and the U.S.S.R., it too shows up clearly in satellite photo- graphs. Detailed fieldwork by the Rus- sian geologists S. D. Khilko, N. A. Flo- rensov and their colleagues shows that left-lateral displacement of a few meters extended for a distance of 370 kilome- ters during the 1905 earthquake. Per- haps the longest and greatest strike-slip fault in the world is the Altyn Tagh fault just north of Tibet. As far as we can determine it was unknown as an active fault, at least outside China, until the satellite pictures became available.

For us the recognition of major strike- slip faults in Asia was an exciting mo- ment because it was immediately clear that there is a fairly simple pattern to them. By combining fault-plane solu- tions of earthquakes in Asia with the observed displacements on the earth's surface following some of the largest earthquakes and with the tectonic fea- tures linked to the faults that are visible in the satellite pictures we have been able to determine the sense of motion associated with the principal faults. In China and Mongolia the strike-slip faults generally run approximately east- west and consistently with left-lateral motion. Faults in the Tien Shan area, and some in or near Mongolia, run north-south or northwest-southeast and consistently with right-lateral motion. This is the pattern one would expect from a squeezing of the region in a di- rection lying between north-south and northeast-southwest, a squeezing that the collision between India and Eurasia would probably cause. Moreover, the predominance of strike-slip motion sug- gests that the northward motion of India is accommodated essentially by materi- al moving out of the way of the imping- ing continents.

In general the clearest, and therefore presumably the most important, faults are left-lateral faults that run east-west. Thus most of the lateral displacement of China out of the way of India would be to the east. This is understandable; ma- terial displaced to the west would have to push against thousands of kilometers of the Eurasian land mass. The motion of China to the east is easily accommodated by its thrusting over the oceanic plates along the margins of the Pacific. Anyone who squeezes a tube of toothpaste experiences a homely analogy to this type of displacement. The thumb and fingers correspond to India and the rest of Eurasia. The closed end of the tube is analogous to the vast continental region of Eurasia that essentially blocks large-scale movement toward the west. The open end of the tube is analogous

to the subduction zones of the western Pacific, with China and Mongolia acting as the toothpaste.

The existence of major strike-slip faults also helps to account for some other apparent peculiarities in the geology of Asia. Although most of Asia appears to be experiencing horizontal compression in a direction between north and northeast, there are two notable exceptions. Lake Baikal occupies a part of the Baikal rift system that is an expression of a northwest-southeast extension. Similarly, Deng Qidong and his

colleagues at the Geological Institute in Peking describe the Shansi graben in eastern China as a rift system created by another extension with the same orientation. We interpret both of these systems as being partly the result of their proximity to strike-slip faults and as being comparable to the tension cracks that develop at the ends of shear cracks.

On this interpretation, after India collided with Eurasia the relative velocity of motion decreased, but India continued to drive into Eurasia at a rate of five centimeters per year. In so doing it

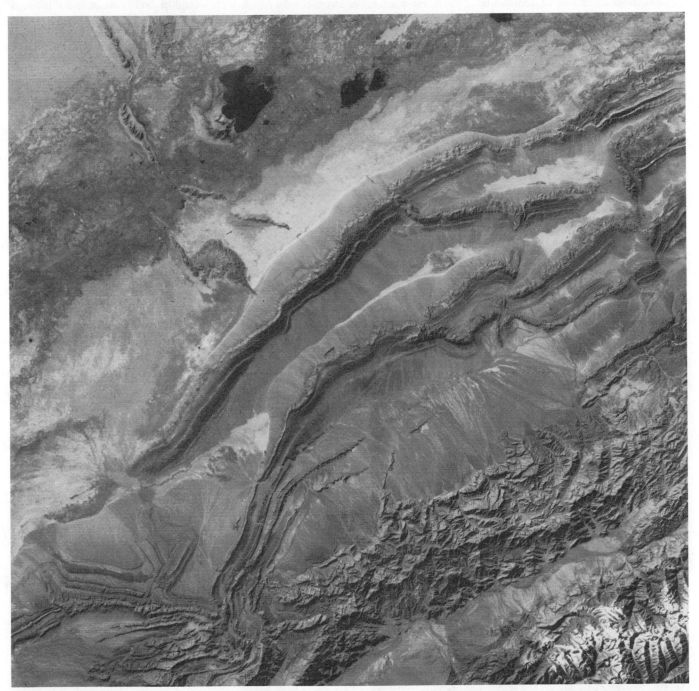

CRUSTAL THICKENING AND SHORTENING appear to have taken place in the region of the Tien Shan range southeast of Lake Balkhash, as is indicated in this ERTS photograph centered on 40.5 degrees north latitude and 78.5 degrees east longitude. The picture shows folded sedimentary formations on the south side of the Tien Shan resulting from thrust faults in the crust under the sediments and dipping to the north under the Tien Shan. Seismic studies show that the crust is 20 to 30 kilometers thicker to the north of the Tien Shan than it is in stabler adjacent areas. This fact suggests that the earth's crust in the eastern portion of the Tien Shan has been shortened, or compressed, by as much as 300 kilometers, presumably because of the northward thrust of India, more than 1,000 kilometers away.

squeezed up mountains in front of it, but more important it pushed material out of its way. The flow of that material may have opened up rift valleys in a direction perpendicular to the direction of shortening, so that in a sense India has wedged Eurasia apart. The overall pattern of deformation is physically similar to that caused by the indentation of a plastic medium by a blunt tool. We think the solutions developed by mechanical engineers for indentation problems may enable us to treat the state of stress in Asia quantitatively and may lead to a better understanding of the large-scale deformation of continents.

The basic unproved assumption in the scheme we have been describing is that large horizontal displacements have actually occurred on the strike-slip faults of Asia. To the best of our knowledge the most thoroughly studied of these faults is the Talasso Fergana fault in the U.S.S.R. There is controversy among Russian geologists over how much displacement has taken place on the fault, but V. S. Burtman suggests that it may amount to 200 to 250 kilometers of right-lateral motion. He suggests, however, that much of the displacement occurred before India collided with Eurasia. He is unable to estimate how much

has taken place since then. As for the other faults, we can only say that most of them are as prominent on the satellite photographs as the San Andreas fault in California is, which has undergone 300 kilometers of right-lateral displacement in the past 25 million years. If the displacements along the highly visible Asian faults are comparable to the displacement along the San Andreas fault, as we suspect, one can conclude that most of the 2,000 kilometers of convergence between India and Eurasia can be attributed to the lateral displacement of China. On the other hand, our hypothesis would be fatally wounded if it could be shown that the displacement along the major faults has amounted to only a few kilometers or at most a few tens of kilometers over the past 40 million years.

In any case it is virtually certain that strike-slip faulting plays a key role in the process of suturing continents together. McKenzie noted some years ago that in the Middle East, the other important region where continents are actively colliding, the motion of the Arabian subcontinent toward Eurasia is forcing part of Turkey to move to the west in a direction perpendicular to that of the converging continental blocks. Moreover, it is clear from detailed studies by Mattauer and his colleagues of portions of some old mountain belts such as those in France and Spain and in Morocco that strike-slip motion was important long after the continents had collided. It seems quite likely that as other ancient mountain belts are studied, evidence for large-scale horizontal movement along strike-slip faults will be found. Such analyses of continental tectonics are clearly not a direct application of plate tectonics to continents. To apply plate tectonics to the deformation of Asia would call for so many plates that the concept's utility would be lost. We suspect that the same will be found to be true for older continental collisions.

At the same time we view the tectonics of Asia as being a direct consequence of plate motions. The earthquakes and great faults of India, China, Mongolia and the U.S.S.R. may be attributed to a simple phenomenon: the northward motion of the Indian subcontinent riding on the India plate toward the Eurasia plate. What is perhaps most interesting about this interpretation is that it indicates that the movement of India caused the deformation of a region more than 3,000 kilometers away. Since the mountains were created by movement along the faults, and since the climate of the region is in turn profoundly influenced by the topography, environmental conditions throughout much of Asia, including the harsh climate of the Himalayas, the Tibetan plateau, the Gobi Desert, Mongolia and parts of China, can also be attributed to a collision that has been in progress for 40 million years.

GOBI-ALTAI EARTHQUAKE OF 1957, which occurred along the Bogdo fault in the Gobi Desert, gave rise to the crustal displacements shown in these photographs made by the geologist V. P. Solonenko. The earthquake, whose magnitude was 7.9 on Richter scale, produced a left-lateral strike-slip motion, which means that far block, as viewed from either side, moved to left. Motion reached 10 meters, among largest strike-slip displacements of recent times.

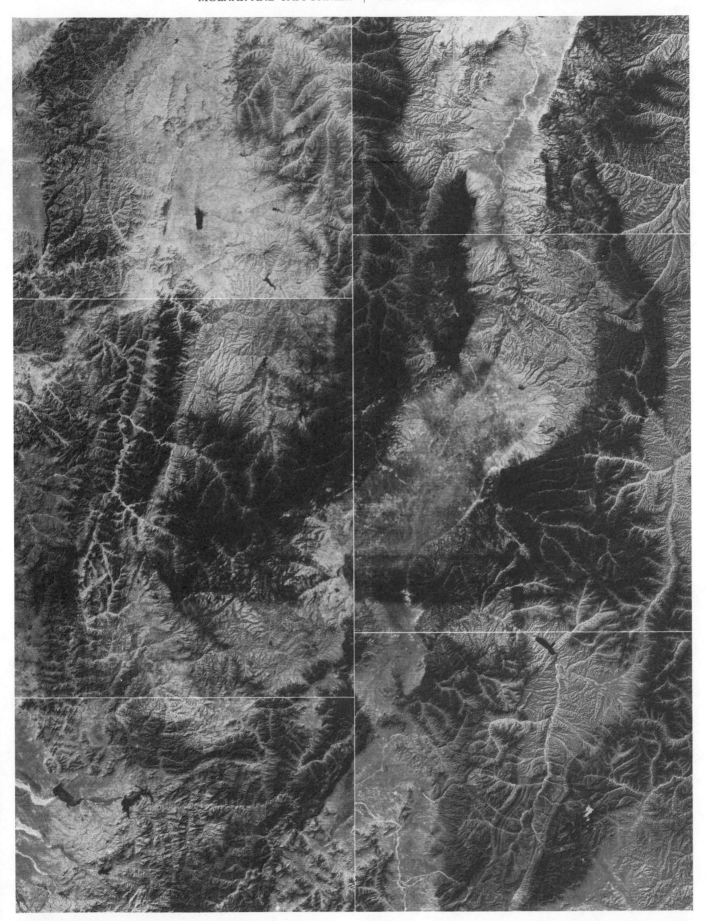

PORTION OF SHANSI GRABEN SYSTEM separates the eastern end of the forbidding Ordos plateau, part of the Gobi Desert, from the fertile, heavily populated valley of the Huang Ho. A graben is a sunken region where crustal blocks are being pulled apart. This mosaic of six ERTS photographs shows the surface features in an area 260 kilometers by 345 kilometers near Taiyuan, about 500 kilometers southwest of Peking. The pictures in the mosaic were made at different times of the year, so that in some of them snow and ice appear on ground and rivers. Shansi graben and the Baikal rift system resemble tension cracks that appear at oblique angles near strike-slip faults.

6

The Fine Structure of the Earth's Interior

by Bruce A. Bolt
March 1973

The waves sent out by earthquakes and nuclear explosions have been studied in detail with new seismometer arrays. They show, among other things, that the core of the earth has a solid kernel

The echoes of an earthquake ring deep into the earth. Depending on the medium through which they pass, the seismic waves are bent, speeded up, slowed down and in the case of some vibrations stopped altogether. When the faint echoes of the earthquake emerge at the earth's surface, they actuate the sensitive pendulum recorders known as seismographs. By correlating the records of seismographs at different locations it is usually a straightforward matter to establish the time and location of the original event, provided that it exceeded a certain magnitude.

During the past decade the worldwide interest in man-made seismic events (nuclear-test explosions) has led to a new generation of sensitive seismographs and the development of seismic arrays that have provided investigators with remarkably effective probes for studying the earth's interior. The delicate traces of the new instruments have enabled seismologists to confirm many of the older hypotheses of the earth's internal structure and have led to the discovery of fascinating new features. Just as radio telescopes have revealed celestial objects that were once invisible, the new generation of seismic instruments has detected fine details of our globe that were once unobservable.

As an example, geophysicists have speculated for years on the nature of the earth's center. Is it a solid or a liquid, and what is its density? The answers have been put on a much firmer basis within the past year. At the earth's center is a solid inner core with a density about 13.5 times the density of water. The radius of this core is some 1,216 kilometers, which makes it a little larger than the moon. The solid inner core is surrounded by a transitional region a little more than 500 kilometers thick. The transitional region in turn is surrounded by a liquid outer core about 1,700 kilometers thick.

Outside the liquid outer core is a mantle of solid rock some 2,900 kilometers thick, which approaches to within 40 kilometers of the earth's surface under the continents and to within 10 kilometers under the oceans. The thin rocky skin surrounding the mantle is the earth's crust. No drill has penetrated the earth's crust deeper than a few kilometers.

The basic technique in deep-earth "prospecting" is to measure the time of travel of a seismic wave from the instant of its generation by an earthquake or explosion to its arrival at a seismographic station. The distance between the seismic event and the station is expressed as the angle (designated delta) formed at the earth's center between radii drawn to these two points. Thus if the earthquake were anywhere on the Equator, it would be 90 degrees "distant" from a recording station at the South Pole. Similarly, if the earthquake were on the Equator in the Andes at 80 degrees west longitude, it would be 90 degrees distant from a recording station on the Equator in Gabon on the west coast of Africa at 10 degrees east longitude.

Earthquakes and explosions generate two types of wave that travel through the interior of the elastic earth. In seismology they are referred to as P (compressional) waves and S (shear) waves. The speed of the waves depends on the density and elastic properties of the rocks through which they pass. S waves are slower than P waves and do not pass through regions that are liquid. A wave that arrives at the seismograph without reflection is designated by a single P or S. If the wave is reflected once at the earth's surface, it is labeled PP or SS. In addition various letters and numbers can be inserted between the two (or more) P's or S's to indicate more specifically the region through which the wave has passed. Thus a compressional wave that passes through the earth's central core is called PKP, or P' for short. The wave that is reflected from the opposite side of the earth is labeled $P'P'$.

During the first half of this century seismologists painstakingly built up the curves of travel time against angular distance for P and S waves using hundreds of specially selected earthquakes from around the world. In 1906 the British seismologist R. D. Oldham first proposed that earthquake waves show that the interior of the earth is not featureless. He explained the pattern of arrival of the dominant P and S earthquake waves by postulating that the earth has a large central kernel. Although the details of his argument proved to be incorrect, Oldham's general conclusion was later verified in many ways. Somewhat stronger evidence of discontinuous features in the earth's structure was discovered in 1909 by Andrija Mohorovičić of Yugoslavia. He found that when he plotted the time-distance curve for P waves from Balkan earthquakes, there was a sharp bend near a distance of about 200 kilometers (an angular distance of about two degrees). Mohorovičić explained the bend by supposing that at a depth of about 50 kilometers there was an abrupt change in the properties of the earth's interior. That discontinuity, which separates the superficial crust from the mantle, was later found to be worldwide.

The detection of this discontinuity below the surface was soon followed by the firm location of the much greater discontinuity, nearly halfway to the earth's center, that had been proposed by Oldham. Seismologists had noted that at distances of up to 100 degrees P waves from earthquakes were recorded with more or less uniform amplitudes but that at distances beyond 100 degrees the amplitudes decayed dramatically. The short-period *P* waves appeared strongly and consistently again only at distances greater than 140 degrees, but at these larger distances they arrived some two minutes later than one would expect on the basis of simply extrapolating their velocity for distances less than 100 degrees.

An explanation that fitted the observed pattern of travel times was worked out numerically in 1913 by Beno Gutenberg of the University of Göttingen. He calculated that at a depth of about 2,900 kilometers the velocity of *P* waves falls precipitously by about 40 percent. This discontinuity marks the structural boundary between the mantle and the core. Seismograms show that *S* waves can travel anywhere above this major boundary, indicating that the mantle material must be rigid. For at least 2,000 kilometers below the core boundary, however, no *S* waves have been observed to propagate. For this reason and others the outer 2,000 kilometers of the core are regarded as being molten.

One might expect that a sharp boundary between the mantle and the core would reflect seismic waves. Indeed, such echoes are clearly observed on seismograms. A direct reflection of *P* waves from the boundary between the mantle and the core is designated *PcP* [see illustration below]. Echoes of the *PcP* type provide a direct means of determining the depth of discontinuities in the deep regions of the earth.

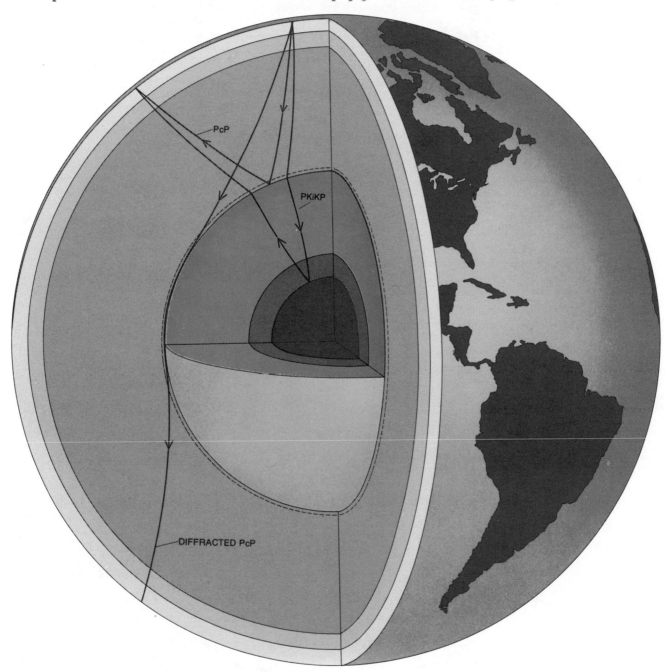

CROSS SECTION OF THE EARTH is based on the most recent seismological evidence. The outer shell consists of a rocky mantle that has structural discontinuities in its upper part and at its lower boundary that are capable of reflecting or modifying earthquake waves. Below the mantle an outer fluid core surrounds a solid kernel at the earth's center; between the two is a transition shell. The paths taken by the three major kinds of earthquake wave are depicted. The waves reflected from the outer liquid core are designated *PcP*; the waves reflected from inner solid core are *PKiKP*; the waves that creep around the liquid core are diffracted *PcP*.

In the 1920's, with the aid of somewhat improved seismographs and more refined earthquake surveillance, it became possible to detect the onset of delayed waves of the *P* type at distances between 110 degrees and 140 degrees. At such distances the waves evidently passed through the core and were of the *PKP* type. Inge Lehmann of Denmark suggested in 1936 that the pattern of travel times for these *PKP* waves could be explained if the core consisted of two regions, an outer one and an inner one. This notion was endorsed by Gutenberg, who now was working with Charles F. Richter at the California Institute of Technology, and independently by Harold Jeffreys of the University of Cambridge. In one of the most impressive sustained studies ever carried through in the physical sciences the two Cal Tech seismologists and Jeffreys (who worked in the early stages with K. E. Bullen) independently computed average velocity distributions for the whole of the earth's interior based on analyses of thousands of seismographic records of *P* and *S* waves. In general these independent solutions agreed as well as the measurements of that time allowed.

With the discovery of the inner core all the major boundaries within the earth had apparently been located. Other discontinuities were suggested from time to time to account for observed discrepancies in seismograms, but by and large

measurements from the available instruments did not have the resolving power to clinch the case for further significant worldwide deep structures. In the period immediately following World War II, however, many seismologists became convinced (particularly from studies of seismic waves of long wavelength traveling around the earth's surface) that the structure of the upper few hundred kilometers of the earth was complex and also that it varied from place to place. The velocities of *P* and *S* waves in many geographical regions inexplicably decreased in a layer below the crust.

Around 1960 observational seismology took a major leap forward. Largely as a consequence of the attempt by several countries to find ways to discriminate between underground nuclear explosions and natural earthquakes, seismology was transformed from a neglected orphan of the physical sciences into a family favorite. A global network of more than 100 standardized sensitive seismographic units was established with U.S. support, and many other earthquake observatories were modernized throughout the world. Arrays of seismographs, comparable to giant radio antennas, were constructed by the U.S., Britain and other countries. In these seismic arrays the seismographs are connected in such a way that microseisms— the random small quivers of the earth—

can be filtered out. One such giant antenna is the Large Aperture Seismic Array (LASA) located near Billings, Mont. It consists of 525 linked seismographs distributed over an area 200 kilometers in diameter [*see illustration on this page*].

In the more than three decades following Lehmann's suggestion that the earth's core might itself have a core, no unequivocal evidence on the nature of the inner core's boundary had ever come to light. The evidence was finally supplied by LASA. In 1970 E. R. Engdahl of the National Oceanographic and Atmospheric Agency, Edward A. Flinn of Teledyne Incorporated and Carl F. Romney of the Air Force Technical Applications Center announced that the Montana array had detected the echoes, designated *PKiKP*, that had bounced steeply back from the boundary of the inner core. The source of the echoes was underground nuclear test explosions in Nevada as well as earthquakes [*see top illustration on opposite page*]. There were two immediate conclusions. First, the inner core has a sharp surface. Second, its radius is within a few kilometers of 1,216 kilometers (a value, incidentally, that I had predicted eight years earlier on the basis of other seismological evidence).

Even more can be done with these remarkable observations. By comparing the relative strengths of the *PcP* and *PKiKP* pulses one can calculate the density of the rocks at the top of the earth's inner core. There are a number of uncertainties in such calculations. The equations assume that the speeds of *P* waves are well determined throughout the earth and that the densities of the materials on each side of the boundary between the mantle and the core are known. Some recent calculations I have made in collaboration with Anthony I. Qamar indicate that the density at the center of the earth cannot be much greater than 14 times the density of water. Earlier estimates had ranged as high as 18 times the density of water. Our value agrees quite well with the density that iron is estimated to have when it is subjected to the pressure that exists at the earth's center.

Let us now turn our attention from the core to the fine structure of the upper part of the earth's mantle, the region just under our feet, so to speak. The presence of great mountain ranges, mid-ocean ridges and deep ocean troughs, together with a considerable amount of geological and geophysical evidence, indicates that large portions of the upper

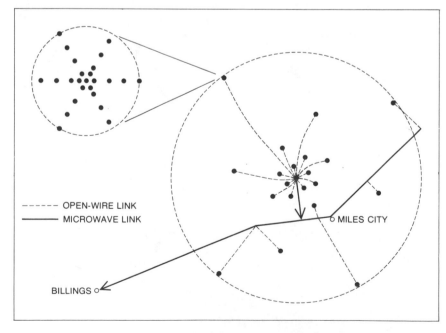

----- OPEN-WIRE LINK
—— MICROWAVE LINK

○ MILES CITY

BILLINGS ○

LARGE-APERTURE SEISMIC ARRAY (LASA) was installed near Billings, Mont., in the mid-1960's by the Department of Defense. One of the largest of more than 100 seismographic stations established throughout the world with U.S. support to detect underground nuclear explosions, it has provided much new knowledge of the earth's interior. LASA consists of 525 linked seismometers grouped in 21 clusters. In each cluster 25 seismometers are arranged as shown at the upper left. The array covers an area 200 kilometers in diameter.

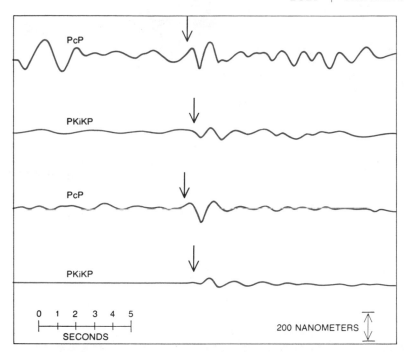

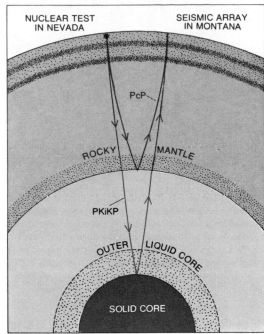

UNDERGROUND NUCLEAR EXPLOSION in Nevada on January 19, 1968 (code-named "Faultless"), produced the traces shown at the left on LASA seismometers. Time proceeds from left to right. The vertical scale shows the magnitude of ground movement involved; 200 nanometers is only half the wavelength of violet light. The *PcP* echoes from the outer core (*diagram at right*) are the closest to a straight-down-and-back path yet reported. Angular distance between explosion and recording instruments was only 11 degrees; this was the angle between lines from two points to the earth's center. *PKiKP* pulses represent echoes from solid inner core.

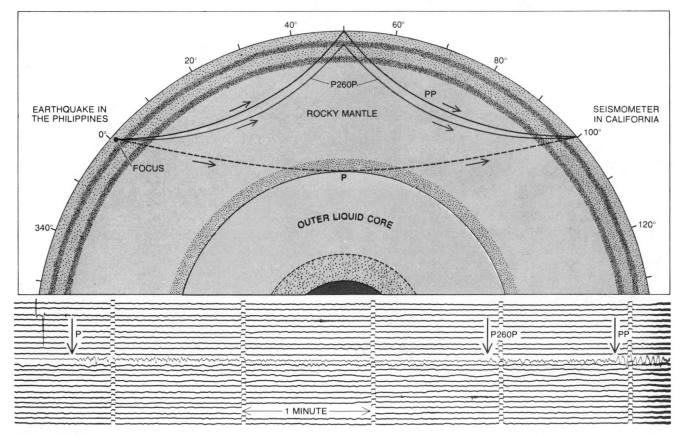

PHILIPPINE EARTHQUAKE generated waves that followed three distinct paths (*top*) before arriving at a high-gain seismograph located in an abandoned mine in the Sierra Nevada. The installation, the Jamestown Station, is operated by the University of California at Berkeley. In the seismograph a spot of light is focused on a rotating drum, producing a series of parallel lines (*bottom*). The line in the Jamestown seismogram that contains three distinct echoes from the Philippine earthquake is shown in color. The *P* wave arrived first, followed some three minutes later by the echo *P260P* that probably bounced off a reflecting surface 260 kilometers deep in the earth under the Pacific Ocean. The reflection from the underside of the ocean floor, *PP*, arrived another minute later.

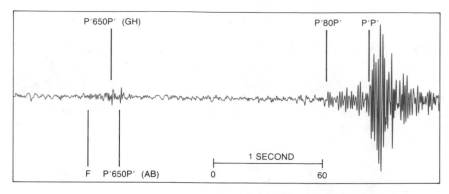

UNDERGROUND NUCLEAR TEST on October 14, 1970, on the Russian island of Novaya Zemlya produced this seismic trace at the Jamestown Station. The large phase $P'P'$ was produced by a compressional wave reflected from the other side of the globe under Antarctica, as depicted in illustration on the opposite page. It was preceded by echo $P'80P'$, evidently reflected from a structure 80 kilometers below surface of Antarctica. Two minutes earlier still record shows doublet $P'650P'$ (GH) and $P'650P'$ (AB), evidently reflected from a layer 650 kilometers below surface of Antarctica. Origin of wave train starting at F is unknown.

part of the earth are in continuous slow motion. Indeed, it now appears that the surface of the earth is divided into six to eight large platelike regions that move with respect to one another. Africa is on one such plate; North America is on another. Dynamic processes operating particularly at the edges of plates would seem to account for much of the topographic relief on the surface of the globe. Most volcanoes are located and most earthquakes occur near the plate margins. One would expect large-scale movements of crustal masses to be reflected in the architecture of at least the upper few hundred kilometers of the earth. One would also expect this variable architecture to give rise to varied patterns on seismograms, and indeed different patterns are observed.

A few years ago, working with Qamar and Mary O'Neill at the University of California at Berkeley, I found a series of unexpected waves on seismograms. Measurements indicated that the waves might have been reflected from the underside of layers located perhaps hundreds of kilometers below the Mohorovičić discontinuity. The waves arrived at seismometers in the Berkeley network as much as 150 seconds before the corresponding waves reflected from the underside of the earth's surface.

We refer to these echoes generically as PdP waves. When we are able to calculate the depth below the surface of the layer from which the waves are reflected, we insert the figure in place of d. Thus a wave reflected from a layer 260 kilometers below the surface is designated $P260P$. A clear example of such an echo is one that was produced by an earthquake on May 22, 1972, near the Philippines and recorded at Jamestown,

Calif., near the Nevada border [*see bottom illustration on preceding page*]. The distance between the focus of the earthquake and California is about 100 degrees.

Such reflections do not always show up on seismograms, and for this reason and others some seismologists have suggested that the wave paths are not as symmetrical as we have proposed. Notwithstanding the reservations of our colleagues, we are satisfied that many observations of PdP waves can be explained in terms of a roughness of the rocky material in the upper part of the earth's mantle that is capable of producing reflections. It is noteworthy that a significant discontinuity in the upper mantle was suggested as far back as 1926 by Perry Byerly of the University of California at Berkeley on the basis of travel times for waves produced by an earthquake in Montana. Byerly's plot of observed travel times against distance showed peculiarities that could be explained if the P waves had encountered some kind of surface at a depth of about 400 kilometers.

Our interpretation has been strengthened by slightly different types of analyses. In 1968 R. D. Adams of the Seismological Observatory in New Zealand and a year later Engdahl and Flinn in the U.S. independently observed small waves that arrived slightly earlier than the usual $P'P'$ echoes. Ordinary waves of the $P'P'$ type make the long journey from the focus of an earthquake to the other side of the earth and are reflected back to a station in the same hemisphere as the earthquake, having passed through the core twice. Adams and Engdahl and Flinn interpreted the precursor waves as $P'P'$ waves that did not quite reach the

opposite surface of the earth but were reflected back from a discontinuity in the upper mantle.

Waves of the $P'P'$ type are particularly useful for probing the earth's structure. Their path is so long that they arrive some 39 minutes after they have begun to be generated by an earthquake. Therefore when they reach the seismograph most of the other waves sent out by the earthquake have already arrived at the observatory and the instrument is quiescent.

A particularly striking example of multiple long-distance reflections was provided by an underground nuclear explosion at the Russian test site in Novaya Zemlya on October 14, 1970. The $P'P'$ waves passed through the earth's core, were reflected under Antarctica and returned to the Northern Hemisphere. In a recording made at Jamestown the main echo $P'P'$ is the most prominent feature on the seismogram [*see illustration on this page*]. About 20 seconds before the onset of the large $P'P'$ reflections a train of much smaller waves begins that can be explained as reflections from the underside of layers located in the 80 kilometers of rock below the surface of Antarctica. These forerunner waves are thus designated $P'80P'$.

As the eye scans the seismogram further from right to left only inconsequential waves are seen for more than a minute and a half; they are minor jiggles continuously produced by the background microseismic noise of the earth. All at once, almost precisely two minutes before the first $P'80P'$ waves, one can see a beautiful doublet: two sharp peaks, separated by a few seconds, that stand out clearly above the background shaking. These sharp pulses agree nicely with the expected travel time of rays reflected by a layer located some 650 kilometers below the surface of Antarctica; hence they are designated $P'650P'$. The presence of a doublet means that there was only a slight variation in the paths of the two rays reflected from the 650-kilometer layer. Presumably one of the rays entered the transition layer, or F layer, that is thought to lie between the inner solid core and the outer liquid core and the other ray did not [*see illustration on opposite page*].

Many other examples have now been reported, particularly by James H. Whitcomb of the California Institute of Technology, of $P'P'$ waves arriving earlier than one would expect if the upper mantle of the earth were uniformly smooth. Most seismologists now agree that these precursor waves at least indicate the existence of a rather sharp boundary at a depth of 650 kilometers and that the boundary is probably a worldwide feature.

It should be noted, however, that the particularly clear seismographic record of the Novaya Zemlya test explosion shows no spikes between the $P'650P'$ waves and the $P'80P'$ waves, as one would expect if there were a sharp reflecting surface at some intermediate depth, such as the 260-kilometer depth I have mentioned in connection with the record produced by the earthquake near

the Philippines. The absence of intermediate spikes indicates either that such shallower discontinuities are not present under Antarctica or that they are not so easily detected by short-period P waves that pass through the earth's outer core because the shallower discontinuities are less sharp than the one at 650 kilometers.

Although the geophysical extent and the precise depth of some of the abrupt changes of structure in the upper mantle remain indefinite, the newly studied class of PdP and $P'dP'$ reflections strongly indicates that sharp variations in rock

properties do exist in the top few hundred kilometers of the earth. That is just the region where, according to the new plate-tectonic view of the earth's crust, there must be a driving mechanism to account for the movements of continents and crustal plates. The mechanism most frequently proposed is the slow flow of viscous hot rock in the form of convection cells immediately below the plates in the upper mantle. It is not obvious how such a flowing region could be reconciled with the observations of sharp structural discontinuities. Geophysicists

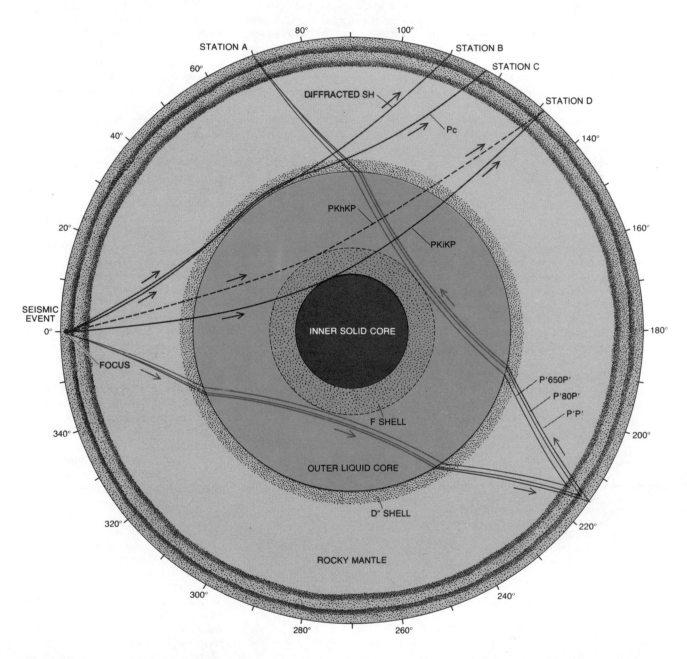

SEISMIC-WAVE PATHS are superposed on a diagram showing the shells that make up the earth's interior. The three nearly parallel rays arriving at Station A represent the paths taken by seismic waves from the Russian underground nuclear test whose seismographic record is shown on the opposite page. The two rays arriving at Station B and Station C represent the paths of diffracted SH waves and Pc waves depicted in the illustration on the next page and in the top illustration on page 277. Studies of such waves clarify the nature of D'' shell. Deeper rays arriving at Station D represent the paths of waves shown in the bottom illustration on page 277.

are now busy examining whether or not convecting rocks can maintain sharp boundaries at certain depths.

Seismological evidence has long existed that suggests some kind of transition region or thin shell at the base of the rocky mantle just above the liquid outer core at a depth of about 2,900 kilometers. This inferred shell was given the label D'' even before much was known about it. Recently fresh light has been thrown on the velocities of P and S waves in the D'' shell, allowing some inferences about its properties.

In the first place modern instruments, unlike early seismographs, do detect clear waves of the short-period P type at distances well beyond 100 degrees. Such observations are surprising, because if the wave speeds remained constant or increased at the bottom of the mantle, the core would act as an opaque screen that would quickly cut off the direct, short-period seismic waves at about 105 degrees. Beyond this distance the seismographic stations would be in the shadow of the earth's core.

The actual situation can be compared to what happens when a shadow is cast on a wall by an opaque object illuminated by the sun. The shadow is not quite sharp because the light waves are diffracted, or bent, by the edge of the object. In the earth some seismic waves get diffracted into the shadow produced by the earth's core. Regardless of the actual velocity of P waves at the base of the mantle, therefore, the shadow region is always dimly illuminated with earthquake waves. It turns out, from work done by Lehmann and by Robert A.

Phinney of Princeton University, as well as from the studies of others, that the strength of the short-period waves actually observed in the shadow of the core is greater than it should be if diffraction in a zone of constant velocity were the only mechanism operating.

Recently I have completed a study of waves designated Pc that have traveled out to more than 118 degrees and have arrived at seismographic observatories in the core shadow zone with substantial strength [see illustration below]. The time of travel and the amplitudes of such waves indicate that they have passed through a shell where the speed of propagation is perceptibly less than the speed that prevails only 100 kilometers or so above the core boundary. Putting the evidence together, the best explanation seems to be that the P-wave velocity drops by a few percent in the D'' shell above the earth's core.

The evidence from the Pc waves does not stand alone. Because the outer core is liquid, certain kinds of shear (S) waves are inhibited from creeping around the core boundary; their energy leaks off into the core in the form of P waves. There is, however, one type of shear wave, called the SH wave, that cannot produce P waves in the liquid core. Its energy gets trapped at the base of the mantle, and thus it can travel great distances. Very large SH pulses have been recorded at distances ranging from 90 degrees to more than 115 degrees from an earthquake [see top illustration on opposite page]. From the measured travel times one can derive the velocity of the shear wave as it travels around the core boundary. Calculations carried out

in recent years by John Cleary of the Australian National University, by Anton L. Hales and J. L. Roberts of the University of Texas at Dallas, by Mansour Niazi, then working at Berkeley, and by others indicate that the S-wave velocity decreases at the base of the mantle in the D'' shell in the same way that the P-wave velocity does.

What kind of earth model can be based on the fine-structure results I have been discussing? The model should also take other facts into account, notably the observed periods of the earth's free oscillations [see "Resonant Vibrations of the Earth," by Frank Press; SCIENTIFIC AMERICAN, November, 1965]. Such a model was computed at Berkeley in 1972. Named CAL 3, it shows the average variation in P-wave and S-wave velocities and in rock density from the surface of the earth to the center [see illustration on page 83].

What does the low velocity in shell D'' mean? One suggestion is that in this narrow shell the mantle rocks become less rigid as they approach the liquid core. Another possibility is that the rock composition changes slightly in the D'' shell. Below the mantle-core boundary the measured physical properties indicate a material that consists mainly of iron. Perhaps there is about 10 percent more iron in the D'' shell than there is in the rocks above it. Such iron enrichment may represent iron that never settled into the core during the formation of the earth or it may represent liquid iron that has diffused outward to form an alloy with the solid mantle.

Currently there is controversy concerning the possible presence of bumps, perhaps 500 kilometers from one side to the other, on the boundary between the mantle and the core. The existence of such bumps has been suggested by Raymond Hide of the British Meteorological Office and by others to explain the variations in the strength of the earth's magnetic and gravity fields as they are measured at different points on the earth's surface. The only way to "X ray" the earth for such fine structure is to use short-period seismic waves that interact with the boundary of the core. It can be calculated, however, that undulations on the mantle-core interface of less than 10 kilometers cannot easily be resolved with waves of the PcP type that return to the surface after a single reflection.

Fortunately, with the advent of sensitive seismographs a method has come into being that has improved resolving power. Some of these seismographs are at sites so quiet that they regularly de-

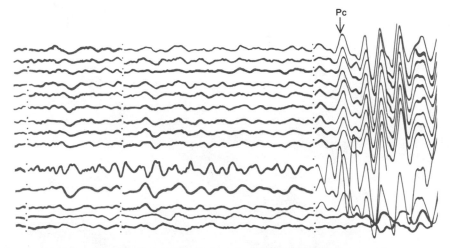

Pc WAVES FROM AN EARTHQUAKE in the South Sandwich Islands were guided with only small loss around the earth's core by the D'' shell. They were recorded at an angular distance of 118 degrees by an array in Uinta Basin of Utah. Here each wavy line is produced by a separate seismometer. Distance between breaks on the bottom line is 10 seconds.

tect waves that have been trapped inside the core and internally reflected four to seven times before emerging and returning to the surface. A wave that has been internally reflected seven times is designated *P7KP* [*see top illustration on next page*]. Such multiple reflections had been predicted but were never seen until the advent of the modern sensitive seismograph and of seismic arrays. It was a great thrill when we scanned along a seismogram made at Jamestown of an explosion on Novaya Zemlya and there —Eureka!—at the travel time predicted for a wave reflected seven times was an unmistakable tiny pulse nestling in the valley of microseismic background noise [*see bottom illustration on next page*].

It is doubtful that we will ever see a natural earthquake generate such a clear example of a *P4KP* or *P7KP* wave. Sharp onsets can be seen when the source is a nuclear test because the energy is released explosively in a way that is simpler than the way energy is released in most earthquakes. In time planned experiments using nuclear explosions, set off in particularly favorable locations and designed to present no hazard to the human environment, should provide still more sensitive probes of the fine detail of the interior of the planet.

What conclusions can be drawn on the nature of the mantle-core boundary from the records of *P4KP* and *P7KP* waves? First, the onset of the waves is quite abrupt. This confirms that the mantle-core boundary is a sharp discontinuity, perhaps extending over no more than two kilometers. Second, the additional distance represented by the extra three legs in the *P7KP* wave reduces its amplitude to about a third of the amplitude of the *P4KP* wave. The small size of the decrease implies that the liquid outer core transmits short-period *P* waves very efficiently indeed.

Finally, if the multiple reflections encountered topographic bumps on the mantle-core boundary more than two kilometers high, the travel times of the waves would be altered enough for the variations to be measurable. By comparing the travel times of many multiply reflected waves at the same seismographic station one should be able to derive the height of the bumps, if any, from the amount of variation in the times. Although studies of this kind are barely a year old, the present indications are that the variation is no greater than it is for waves that do not bounce from the core boundary. Therefore one can say tentatively that if there are topographic undulations, either their height is less than a few kilometers or, if their height is sig-

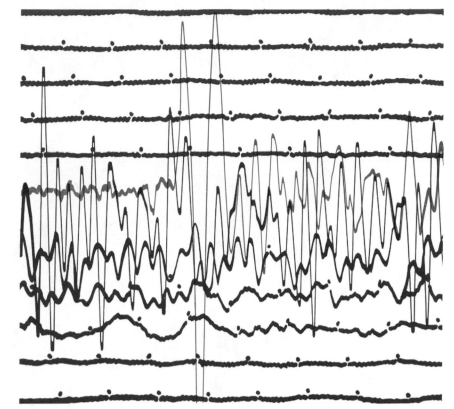

HORIZONTAL SHEAR WAVE of the type designated *SH* was produced by a major earthquake in Iran on August 31, 1968. It was recorded at a distance of nearly 100 degrees by a seismograph at Port Hardy in Canada that responds only to horizontal ground motion. The large *SH* pulse sent pen flying across recording drum, crossing traces made earlier and later.

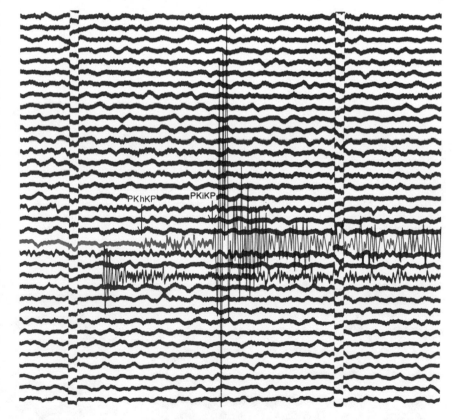

WAVES THAT PENETRATED EARTH'S CORE were created by a deep earthquake near Java and recorded at a distance of 132 degrees by a seismograph at Golden, Colo. The large-amplitude wave *PKiKP* was refracted through the earth's solid inner core. It was preceded 17 seconds earlier by smaller *PKhKP* waves, which were probably reflected from *F* layer.

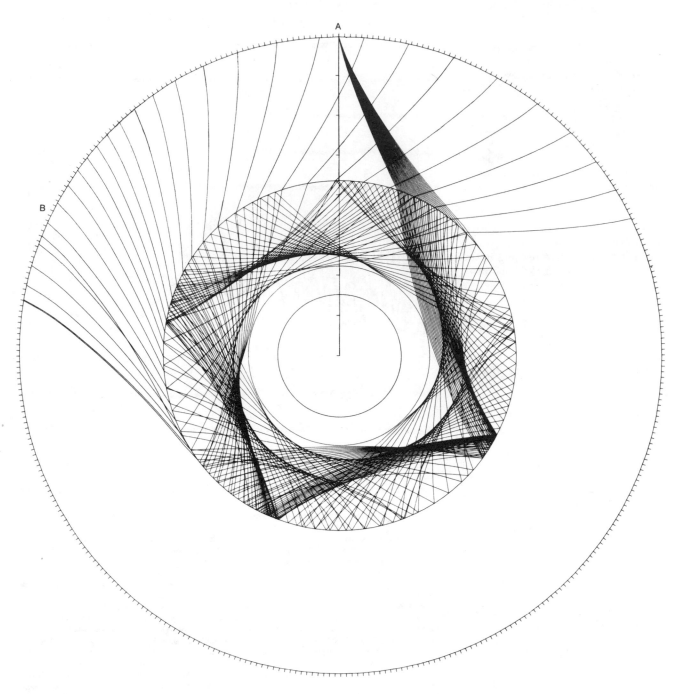

MULTIPLE REFLECTIONS can result from *P* waves that get trapped inside the earth's liquid outer core. This computer plot depicts the paths of waves, generated by a seismic event at *A*, that have bounced inside the core seven times before reaching the surface, for example a station at *B*. Such waves are assigned a *K* for each bounce, hence the waves shown would be designated *PKKKKKKKP*, or *P7KP*. Computer program that produced ray paths was devised by C. Chapman for seismological average earth model called CAL 3.

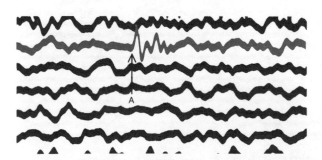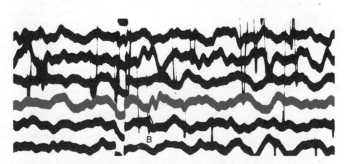

FAINT PULSE OF *P7KP* ECHO can be seen at the right (*B*) in this seismogram made at Jamestown of an underground explosion on Novaya Zemlya in 1970. The stronger *P4KP* pulse, labeled *A* at the left, was recorded about 20 minutes earlier; it represents a wave that was reflected only four times inside the core of the earth. Such evidence indicates that the boundary of the core is rather smooth.

nificantly higher than that, they are few in number.

Let us now focus our attention still deeper in the earth. Even before 1940 the better-equipped seismological observatories reported tiny but clear *P*-wave precursors about 10 seconds before the main onset of the core waves (*PKP*) at distances of between 130 and 140 degrees. Quite unmistakable observations of such precursor waves were reported on seismograms from the nuclear explosions at Bikini Atoll in 1954. A recent typical precursor wave was generated by an earthquake near Java and recorded 132 degrees away at a seismic station in Golden, Colo. The precursor waves arrived 17 seconds before the onset of the much stronger *PKP* wave [*see bottom illustration on page 81*].

One straightforward explanation for these precursors occurred to me in 1962. Perhaps the wave velocities around the inner core were significantly different from the values that were accepted at the time. I suggested that between the inner and the outer core there might be a transition shell, designated *F*, that has a small jump in velocity at its upper boundary. After working out many mathematical models of the earth's core I found that, although other explanations are possible, the hypothesis of an *F* shell predicts travel times in reasonable agreement with all the *PKP* observations, including the precursor waves. Independent studies by Adams and M. J. Randall at the Seismological Observatory in New Zealand soon confirmed the general results I had obtained. The precursor waves reflected from the surface of the *F* shell were named *PKhKP*. Just in the past year an alternative explanation of some of the precursor waves has been put forward by R. A. Haddon of the University of Sydney. He suggests that they are *PKP* waves scattered by bumps on the mantle-core boundary.

There were several fruits of my proposal of 1962. Revised travel times for the core waves, based on the *F*-shell model, not only helped other seismologists to find more examples of the small *PKhKP* precursors but also helped in the search for the reflections (*PKiKP*) from the boundary of the inner core. It was these latter reflections that led to the conclusions that the inner core has a sharp surface and that its radius is about 1,216 kilometers.

An important by-product of the 1962 core-velocity estimates concerns the question of whether the inner core is solid or not. The notion that the inner core might be solid was put forward in 1940 by Francis Birch of Harvard Uni-

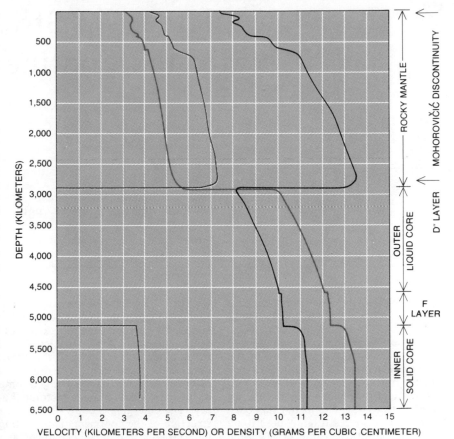

SEISMOLOGICAL AVERAGE EARTH MODEL (CAL 3) was computed last summer in author's laboratory at the University of California at Berkeley. The model, which takes all available seismological information into account, is defined by the three curves. The colored curve shows the variation in the earth's density with depth. The solid black curve shows the average velocity of *P* waves. The gray curve shows the average velocity of *S* waves. Since *S* waves are not propagated through liquids, the gray curve is interrupted by the earth's outer liquid core. The energy in an *S* wave, however, can be transmitted as a *P* wave through the liquid core and then reconverted into an *S* wave for transmission through the solid inner core. Direct evidence for a solid inner core has only been obtained in the past year.

versity. Now, if the inner core is rigid, it should transmit not only compressional waves but also shear waves, giving rise to waves, designated *PKJKP*, observable under specially favorable conditions at the earth's surface.

Working at the University of Sydney with the travel times of *P* waves and the wave velocities then assumed for the core, Bullen made estimates in 1950 of the travel times for *PKJKP* waves. There followed an enthusiastic but unsuccessful search for them. Seismologists at observatories around the world scrutinized seismograms to see if there were unexplained wiggles near the predicted times for *PKJKP* waves. In my own work on the core I estimated in 1964 an S-wave velocity in the inner core of about 3.7 kilometers per second. As a result the search for *PKJKP* waves was shifted to arrival times several minutes later than the ones that had first been tried.

Early last year Bruce Julian, David Davies and Robert Sheppard of the Mas-

sachusetts Institute of Technology provisionally announced that the large-aperture seismic array in Montana had recorded *PKJKP* waves close to the arrival time predicted by the 1962 model. If this announcement is confirmed, a long-sought key to the nature of the earth's kernel will have been found.

What further consequences follow from our new knowledge of the planet's fine structure? Many could be mentioned. For example, since wave speeds depend on rock densities, the revised *P* and *S* velocities in the earth make it possible to estimate the density variation with greater assurance, as incorporated in the CAL 3 model. Geochemists who try to describe the composition of the materials in the various shells will be aided by the more precise determinations of structural boundaries and density. Ultimately the entire body of geophysical and geochemical knowledge must be incorporated into a satisfying account of the evolution of the earth.

7

The Deep Structure
of the Continents

by Thomas H. Jordan
January 1979

The oldest parts of the continents appear to have deep root zones that travel along with the continents as the tectonic plates move. The zones may be chemically distinct from the surrounding rock

Within the past two decades earth scientists have assembled vast amounts of geological, geophysical and geochemical evidence to develop a new picture of the history, structure and dynamic behavior of the outer layers of the earth, informed and rationalized by the theory of plate tectonics. The theory has been particularly successful in describing how the basaltic oceanic crust, some seven kilometers thick, is continually produced at the crests of mid-ocean ridges by the rise of molten magma from the mantle below, how it moves across the surface at the rate of centimeters per year as the top layer of huge, rigid plates 10 times as thick and how it is consumed back into the mantle in subduction zones at the deep ocean trenches. The theory has been less successful in explaining the formation and structure of the continents.

The continental crust, which averages 35 kilometers in thickness, is lighter than the oceanic crust and richer in silicon and potassium. In plate-tectonic theory the continental crust is considered to be a buoyant product of melting and remelting that has accreted in the course of sea-floor cycling over long periods of time; continental drift, the well-documented movement of the continents across the earth's surface, has been explained simply as the passive rafting of this light crust as part of a moving plate. Because the continental crust is buoyant it cannot be mixed back into the mantle in large amounts by tectonic processes such as subduction; it tends instead to float on the denser mantle somewhat like a slag on molten iron. Except for this distinction, plate tectonics postulates no essential differences between the motions of the continents and of the ocean floor, or between the subcrustal structures of the continental plates and of the oceanic plates.

Seismological data, however, reveal substantial contrasts between continental and oceanic structures extending well below the base of the crust. It appears that under the oldest parts of the continents there are deep root zones several hundreds of kilometers thick, which travel along coherently with the continents as the plates move. The discovery of these root zones has challenged some of the basic tenets of plate-tectonic theory. The emerging picture of the deep structure of the continents is providing new insights into the mechanical and chemical processes that control continental evolution and tectonics.

Cratonic Stability

The oldest rocks of the continental crust are found in the basement complex of the continental shields and the continental platforms. The shields are extensive uplifted areas that are essentially bare of recent sedimentary deposits. The platforms are broad, shallow depressions of the basement complex filled by nearly flat-lying sedimentary rocks. Together the shields and platforms constitute the cratons, the stable blocks that are the nuclei of present-day continental masses. Although most of the basement rocks exposed on the shields or buried under platform sediments have been metamorphosed during ancient episodes of orogeny, or mountain building, they have remained undisturbed for very long periods, typically a billion years or more.

The continents comprise more than their stable nuclear cratons, however. In contrast to the cratons are the continents' modern orogenic zones, huge areas that have been pervasively deformed quite recently by tectonic activity resulting from the convergence of two opposing plates. The crust in an active orogenic zone does not behave rigidly; plate tectonics fails to describe its motions. The contrast between cratons and orogenic zones is seen today in southern Asia, which is being deformed by the collision of the northward-driving Indian and Arabian cratons with the Eurasian continent. This violent collision has crumpled Asia to produce a great mountain belt extending from the Anatolian peninsula across the Middle East, the southern U.S.S.R. and most of China to the marginal seas of the Pacific Ocean [see "The Collision between India and Eurasia," by Peter Molnar and Paul Tapponnier; SCIENTIFIC AMERICAN Offprint 923].

A remarkable aspect of this ongoing orogenic event is that the Indian craton continues to drive headlong into Asia at the rate of five centimeters per year, upthrusting high mountains over an area of 10 million square kilometers, without itself deforming. To understand the dynamics of this perplexing tectonic behavior, to account for the violent deformation of Asia and the stolid persistence of an undeformed Indian subcontinent, is a challenging problem for geophysics. Any valid model of continental structure and tectonic behavior must account for this contrast, and it must be based on information from the depths of the mantle, where the forces that have shaped the continents originate.

Continental tectonic behavior is largely controlled by the mechanical structure of the lithosphere, which is functionally defined as the earth's strong outer shell, composed of the crust and the upper part of the mantle. The lithosphere endows the plates with their rigidity. Studies of the earth's gravitational field and of the earth's response to the rapid loading and unloading of its surface by large masses such as glacial ice sheets and mountains suggest that the lithosphere has an average effective thickness of about 100 kilometers. Below it is the asthenosphere, or weak layer, a region of the mantle where even small stresses cause material to flow. According to the plate-tectonic model the lithosphere constitutes the plates and the plates move with respect to the asthenosphere, within which the shear strains associated with plate motion are concentrated.

The strength of the lithosphere is derived largely from the upper layer of the mantle, and the strength of the mantle material varies with temperature. Measurements of thermal gradients near the surface show that temperatures within the lithosphere increase rapidly with depth and that the rates of increase vary widely with geographical location. Laboratory studies and evidence from the field indicate that at a given temperature crustal rocks, ranging in composition from basalts to granites, are significantly less strong than peridotites, the major rocks of the mantle. Because the oceanic crust is thin, mantle peridotites are present at shallow depths, where the temperatures are low. It is apparently this near-surface layer of cold peridotite that gives the oceanic lithosphere most of its strength to resist tectonic deformation. The continental crust, in contrast, is thick, and so the temperatures at its base are higher than those at the base of the oceanic crust. The uppermost mantle under the continents is therefore weaker than its oceanic counterpart, and that weakness seems to explain, at least in part, why the orogenic zones are so easily deformed.

If that is the case, why have the cratons remained stable and undeformed over such great spans of geological history? The answer is closely linked to the problem of deep continental structure. One can argue that although the temperatures at the base of the continental crust generally exceed those at the base of the oceanic crust, the thermal gradient is almost always lower (the temperature increases more slowly with depth) in the cratonic mantle: the temperatures characteristic of flowing asthenosphere are not reached under the cratons until depths greater than those at which they are reached under the oceans. Therefore the integrated temperature of the upper mantle—and hence the effective thickness of the lithosphere—might be greater under the cratons than under the oceans or the orogenic zones. Geophysicists have presumed that the increased lithospheric thickness caused by lower temperatures at greater depths explains cratonic stability.

As we shall see, it is not that simple. Recent work indicates that in addition to the thermal contrasts there are chemical differences within the cratonic roots of the upper mantle that are also important in regulating continental tectonics. Understanding the new and still controversial model of the continental upper mantle that incorporates these compositional differences requires a detailed exploration of the deep structure of the continents.

The Earth's Elastic Structure

Most of what is known of structure within the earth's interior has been revealed by seismology. When powerful sources of elastic waves—earthquakes and large explosions—"illuminate" the interior, the images recorded by seismographs indicate spatial variations in the earth's elastic properties. The elastic response to a seismic source usually is adequately described by three parameters: the density, the speed of compressional waves (whose motion is polarized along the wave path) and the speed of shear waves (whose motion is polarized transverse to the path). Each elastic parameter is a function of pressure, temperature and chemical composition.

For most seismological purposes the earth can be assumed to be a spherically symmetric structure whose pressure, temperature and composition (and hence elastic parameters) vary only with distance from its center. That is the assumption, for example, in the computer algorithms that locate large earthquakes by fitting the observed arrival times of

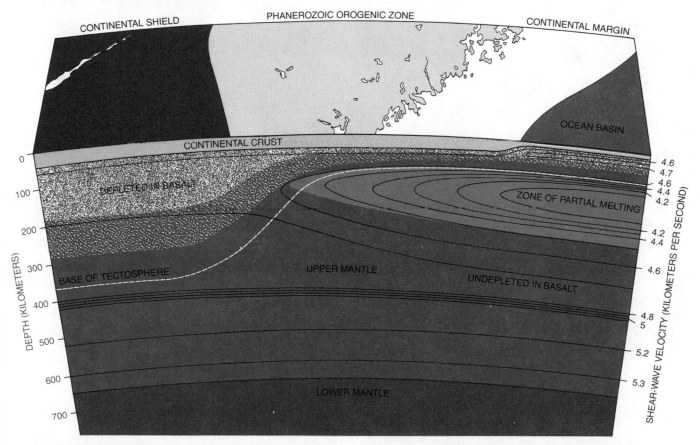

HYPOTHETICAL MODEL of the subsurface transition zone under a typical continental margin shows some of the features that might characterize the earth's upper mantle along a section extending, say, from the Canadian shield (*left*) to the Atlantic Ocean basin (*right*), crossing the New England coast in the vicinity of Portland, Me. The large lateral variations in seismic shear-wave velocities observed in recent measurements are indicated by the black contours. According to the author's basalt-depletion hypothesis, the earth's upper mantle is composed of a mineral assemblage called peridotite (*solid color*), which under the continents is depleted in certain basaltic components; the density of the white dots signifies the degree of depletion. The broken white line represents the approximate minimum depth to the base of the tectosphere, term adopted by the author to define the volume of tectonic plates. Zone of partial melting lies under oceanic crust.

seismic waves. Once an earthquake has been located in space and time, the seismic-wave travel time to any station can be computed. If the earth were really spherically symmetric, that is, if its elastic properties varied only with radius, these travel times would depend only on the great-circle distance between source and receiver and not on their specific coordinates. In fact, however, small but significant geographical variations are observed in the travel times to various stations, typically with magnitudes of less than 1 percent of the total travel time. Analysis of these small regional differences indicates that structures associated with the cratons persist down to at least several hundred kilometers, that is, to depths an order of magnitude greater than the thickness of the continental crust.

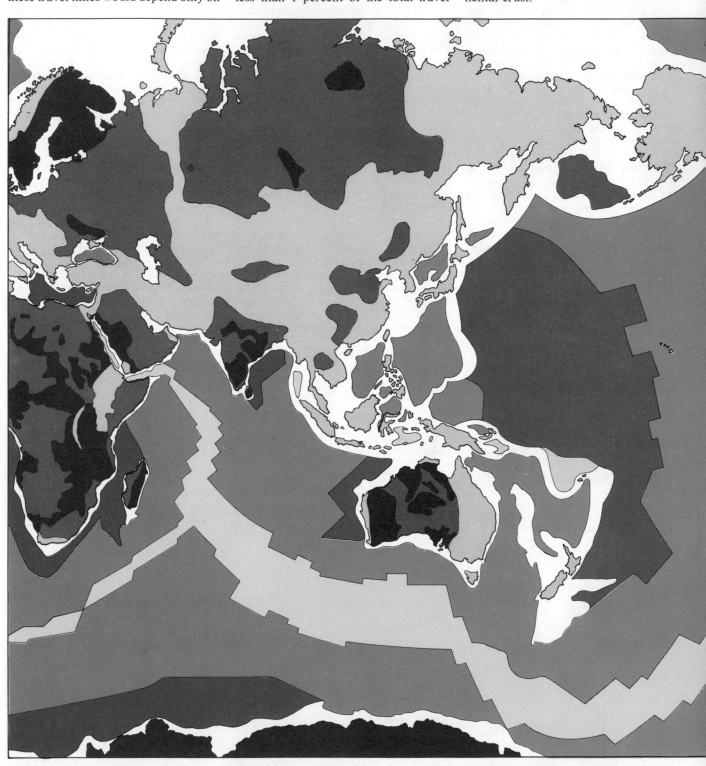

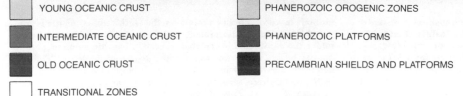

YOUNG OCEANIC CRUST

INTERMEDIATE OCEANIC CRUST

OLD OCEANIC CRUST

TRANSITIONAL ZONES

PHANEROZOIC OROGENIC ZONES

PHANEROZOIC PLATFORMS

PRECAMBRIAN SHIELDS AND PLATFORMS

EARTH'S CRUST IS DIVIDED into seven major types of rock according to the tectonic system of classification presented in the map on these two pages. In general colored areas and white areas are below sea level and gray areas are above sea level (*see key at bottom left*). The lightest color is used to denote regions of oceanic crust that are less than 25 million years old; the intermediate color repre-

The seismological study of continent-ocean heterogeneity is hampered by the dearth of seismic recording stations in the ocean basins. To investigate the upper mantle in regions that lack seismic stations seismologists generally depend on surface waves, whose energy is trapped near the surface and whose travel times depend on the properties of the crust and upper mantle along the entire path from source to receiver. The propagation speeds of the surface waves' low-frequency components are more sensitive to deep structure than those of the higher frequencies, and from measurements of this frequency dependence the variations of the upper mantle's elastic properties can be estimated for both continental and oceanic paths. It was surface-wave data collected in the 1920's by the pioneering seismologists Beno Gutenberg and Robert Stoneley that showed the crust under the oceans to be much thinner than the crust under the continents, but systematic study of large-scale lateral variations below the crust did not begin until the late 1950's and early 1960's, when the first global networks of standardized broad-bandwidth seismometers were installed.

Surface-wave data from these networks quickly convinced seismologists that the structural differences between continents and oceans must extend far below the crust. Throughout most of the crust and mantle the elastic parameters (density and wave speeds) increase with depth because the pressure increases and squeezes the rock into tighter, more rigid structures. Under the ocean basins and most active orogenic zones, however, the shear-wave speed was found to decrease sharply with depth in a transition region that begins about 50 or 100 kilometers below the surface and forms a "low-velocity zone" approximately 100 kilometers thick. Under the cratons, on the other hand, this low-velocity zone is absent or lies deeper and is less prominent.

These seismological findings accorded well with the accepted models of the mantle's thermal structure. Wave speeds generally vary inversely with temperature; a small amount of melting results in a dramatic decrease in velocity. Laboratory studies and theoretical considerations indicated that the upper mantle's low-velocity zone is probably caused by a small amount (1 percent or less) of melting in mantle peridotites owing to the high temperatures about 100 kilometers below the oceans and the orogenic zones. Because the thermal gradients under the cratons are more gradual, the peridotites there melt only at greater depths, if at all.

This structural picture of the upper mantle, which emerged in the early 1960's, also made considerable sense in terms of the rapidly developing theories of sea-floor spreading and plate tectonics. Seismologists were quick to identify the easily deformed asthenosphere with the partially molten material in the low-velocity zone and to identify the rigid lithosphere with the cooler material above the low-velocity zone. It appeared that by mapping the depth to the low-velocity zone seismologists were actually mapping the geographical variations in plate thickness. A corollary followed from the model, however: Any structural variations below the top of the low-velocity zone should not be coherent with the positions of the conti-

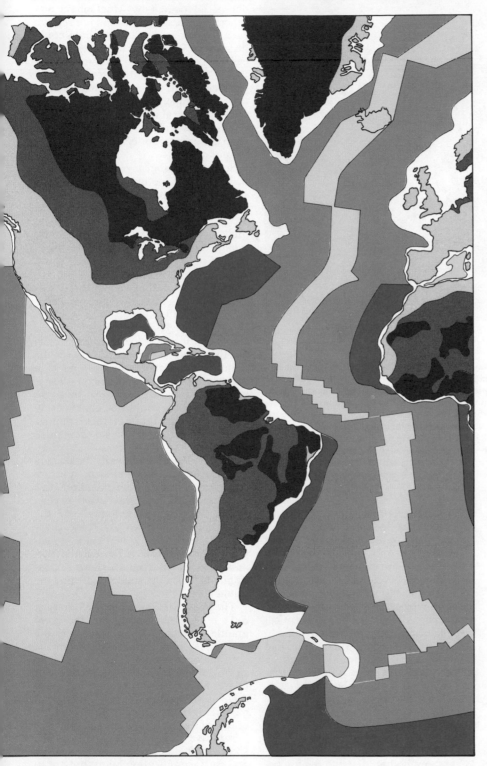

sents regions of oceanic crust that are between 25 and 100 million years old, and the darkest color corresponds to regions of oceanic crust that are more than 100 million years old. The white areas are regions of transitional or submerged continental crust, including the continental margins, the island arcs and the oceanic plateaus. The lightest gray denotes regions of continental crust that have been affected by orogenic (mountain-building) activity in the Phanerozoic era, that is, within the past 600 million years or so. The intermediate gray represents stable continental platforms with a history of sedimentary deposition in the Phanerozoic era, and the darkest gray corresponds to stable continental shields and platforms with no accumulated sediments since the Precambrian era, which came to an end approximately 600 million years ago.

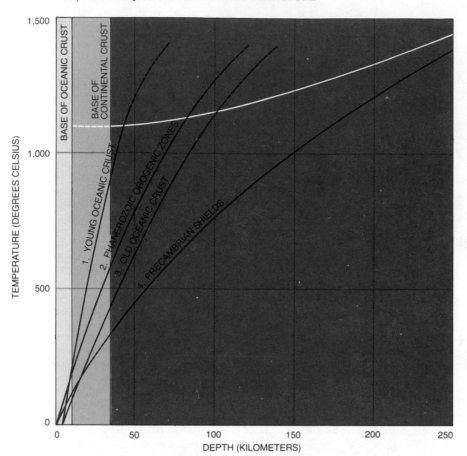

TEMPERATURE of the crust and the upper mantle varies systematically according to the tectonic classification of the crustal rock. The temperature rises most rapidly with depth under the young oceanic regions (*curve 1*) and least rapidly under the Precambrian continental shields (*curve 4*). These variations in the temperature-depth relation appear to be responsible for the observed geographical differences in the velocities of seismic waves. Under most tectonic regions the temperature curves rise rapidly enough to intersect the threshold at which the mantle begins to melt (*white curve*), causing the elasticity of the mantle to decrease suddenly at a certain depth and thereby creating a seismic low-velocity zone. Under the continental shields, however, the temperature curve rises more gradually and melting may not occur. This conclusion is consistent with the view that under shields seismic low-velocity zone may be absent.

nents; plate motions should continually be rearranging the relative locations by moving the lithospheric material with respect to the asthenosphere.

The Depth Issue

From that point of view the seismic evidence for consistently higher shear-wave speeds under the cratons at depths greater than 200 kilometers, below the lithosphere and below the depth of the oceanic low-velocity zone, was disturbing. In 1963 Gordon J. F. MacDonald of the University of California at Los Angeles reviewed the problem of deep continental structure. His models based on the surface-wave data and on information about the mantle's thermal state implied that the lower temperatures and higher wave speeds characteristic of the cratons at shallow depths must extend to depths on the order of 500 kilometers. MacDonald was convinced that such deep structure was incompatible with the notion of a mobile, convecting mantle, and so he concluded that the rela-

tively rapid motions of the continents required by sea-floor spreading and continental drift were improbable, if not impossible.

MacDonald's arguments against drift could not, however, withstand the tide of new information that swept across the earth sciences in the ensuing decade, persuading all but the most skeptical that plate motions are indeed real. In time the data on which MacDonald had based his thesis were reinterpreted by advocates of this "new global tectonics." In 1970 J. G. Sclater and Jean Francheteau, working at the Scripps Institution of Oceanography, formulated thermal models of the oceanic and continental lithospheres that were consistent with plate-tectonic hypotheses. They showed that the differences between continental and oceanic thermal profiles could be confined above a depth of 200 kilometers, rather than extending to 500 kilometers or more as MacDonald had argued, and they proposed a model with an oceanic lithosphere approximately 100 kilometers thick and a

200-kilometer continental lithosphere. Their model received support from Adam M. Dziewonski of Harvard University in 1971. He demonstrated that even if the surface-wave data did allow structural differences between continents and oceans to exist at the great depths advocated by MacDonald and others, the data certainly did not require such variations. In fact, the data could be satisfied by models where the significant structural contrasts were concentrated above 200 kilometers, a value consistent with the conclusions of Sclater and Francheteau and compatible with the general idea of the lithospheric plates. The evidence for MacDonald's deep continental roots appeared to have evaporated.

Some seismologists remained dissatisfied with this picture of upper-mantle structure, however. I. Selwyn Sacks of the Carnegie Institution of Washington analyzed data from very deep earthquakes on the west coast of South America and concluded that the thickness of the plate constituting the South American craton must exceed 300 kilometers. A similar conclusion was reached by Shelton Alexander of Pennsylvania State University, whose analysis of surface-wave dispersion across stable continental blocks led him to argue that structural contrasts between the cratons and ocean basins extend to a depth of at least 400 kilometers. Dziewonski had shown, however, that surface-wave data could not conclusively distinguish structures at these depths, and so most geophysicists remained skeptical.

The evidence that finally confirmed the idea that continental structures are significantly thicker than 200 kilometers came not from the surface-wave experiments but from observations of the more unusual ScS waves. These shear waves travel nearly vertically downward from the source (an earthquake) and are reflected from the sharp discontinuity between the solid mantle and the liquid core.

The phase that is denoted ScS_1 (or just ScS) is reflected once from the core-mantle boundary to the receiver; the phase ScS_2 is reflected twice from the core and once from the free surface, and so on for the higher-order phases. ScS phases have many properties that make them useful for the study of upper-mantle heterogeneity. For example, the travel-time difference between ScS_2 and ScS is sensitive to variations in shear-wave speed in the upper mantle under the ScS_2 surface-reflection point, but not to variations near the source and receiver, where ScS_2 and ScS follow almost the same paths. By varying the recording site (and thus the location of the surface-reflection points of the ScS phases) one can therefore detect lateral heterogeneity in regions where seismic stations are either sparse or absent, such as the deep ocean basins.

In 1975 and 1976, first at Princeton University and then at the Scripps Institution of Oceanography, Stuart A. Sipkin and I published two studies of global velocity variations in the upper mantle based on *ScS* phases. We found that the average travel times of shear waves moving vertically through the mantle and crust under the ocean basins are about four seconds greater than the corresponding travel times for the cratons. Since the oceanic low-velocity zone was known from the surface-wave data to be more accentuated than its continental counterpart, the mere fact that the shear waves moved more slowly under the ocean came as no surprise. It was the large magnitude of the observed difference that was unexpected, because the existing models of structural contrasts between the continents and the ocean predicted much smaller differences; for example, the vertical shear-wave travel times that were computed from Dziewonski's surface-wave models of cratonic and oceanic structures differed by less than one second rather than by four seconds.

These large *ScS* travel-time variations, combined with the surface-wave data, require that significant structural contrasts between oceans and cratons persist to depths that certainly exceed 200 kilometers and probably exceed 400. The surface waves are most sensitive to the variations in elastic parameters near the surface, whereas the *ScS* phases sample the upper mantle almost uniformly at all depths. If one attempts to construct cratonic and oceanic models in which the large contrasts in shear-wave speed demanded by the *ScS* data are concentrated above 200 kilometers, then the surface-wave dispersion curves calculated from the models invar' bly conflict with the dispersion curves nat are actually observed. Indeed, it can be shown with a fair amount of rigor that no reasonable model subject to this depth constraint adequately satisfies both the *ScS* data and the surface-wave data. It appears that only models with significant variations at or below a depth of 400 kilometers can successfully explain both sets of data. Such models can be constructed [*see illustration on page 85*], but they are not unique. Experiments aimed at refining the models are

in progress in my laboratory and elsewhere, and more definitive results should be forthcoming.

Continental Tectosphere

If at this stage one cannot specify with much certainty how the contrasts between continental and oceanic elastic structures vary with depth or even to what maximum depth they extend, it is nonetheless clear that contrasts in shear-wave speed persist considerably below a depth of 200 kilometers. This fact alone forces one to reconsider a concept that has been central to the plate-tectonic model: the notion that the lithosphere everywhere constitutes the plates.

Lithospheric material is by definition strong; it can support the shearing forces induced by surface loads such as mountains for millions of years without much permanent deformation. The determination of lithospheric thickness from observations of loading phenomena and the gravity field was first undertaken by geophysicists in the early part of this century, and modern refinement and reanalysis of the basic data have confirmed one of their original conclu-

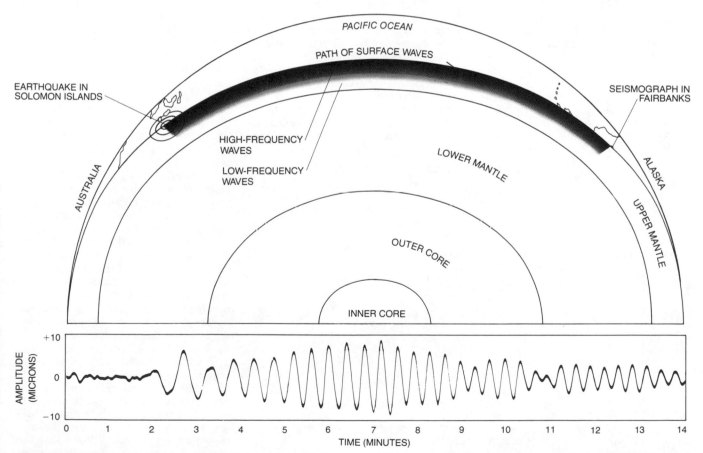

SURFACE WAVES, seismic waves whose energy is trapped near the earth's surface, provide seismologists with a great deal of information about variations in the elastic structure of the upper mantle. The idealized drawing at the top, for example, shows the path of a surface-wave train recorded at a station in Fairbanks, Alaska, following an earthquake in the Solomon Islands in June, 1970. The portion of the actual seismogram reproduced at the bottom shows a characteristic trace made by a Rayleigh wave, a type of seismic wave whose motion is elliptical in the plane of propagation. (Only the vertical component of the wave's motion is represented here.) Such surface waves are said to be dispersive, that is, their speeds of propagation depend on their frequency. Lower-frequency components (*color*) are more sensitive to structural variations at greater depths in the mantle than the higher-frequency components (*gray*), and they generally travel faster.

sions: Under even the most stable cratons such as the Fennoscandian and Canadian shields the effective thickness of the lithosphere is less than 200 kilometers and probably not much more than 100. In fact, the average thickness of the cratonic lithosphere seems to be not much greater than that of old oceanic lithosphere; for example, R. I. Walcott of the Canadian Department of Energy, Mines and Resources has estimated that the lithosphere in Canada, presumably representative of the stable continental areas, is about 110 kilometers thick, whereas the lithospheric thickness under Hawaii, and probably elsewhere in the older oceanic basins, is greater than 75 kilometers. The material below these depths is part of the asthenosphere; when it is acted on by even small shearing forces, it quickly deforms.

The seismic data, in other words, imply that the subcratonic regions of anomalously high shear-wave speeds extend below the lithosphere and into the asthenosphere. As a matter of fact the deep lateral variations in elastic parameters turn out to be remarkably coherent with all the crustal-age classifications shown in the illustration on pages 94 and 95. Two explanations are possible: either the continental deep structures characterized by high shear-wave speeds are more or less statically coupled to the overlying continents—are actually parts of the continental plates—or they represent thermal variations that are somehow dynamically maintained by flows of mass in a convecting mantle. For the latter mechanism to be viable essentially all cratons would have to be the sites of a downward flow of cold

material (because low temperatures are necessary to explain the high shear-wave speeds); such flows would have obvious effects on the earth's gravity field, however, and no such effects are observed. Moreover, a mass-flow hypothesis imposes severe geometrical restrictions on the flow field that appear to be inconsistent with other constraints on mantle dynamics. It seems, therefore, that deep continental structures cannot be explained as features that are dynamically maintained by convection.

One is therefore forced to conclude that these deep structures do indeed constitute the lower portions of the continental plates and that they have been translating coherently with the crust for hundreds or even thousands of millions of years, in spite of the fact that the material forming these deep structures is part of the asthenosphere. Faced with this situation some geophysicists have preferred simply to redefine the lithosphere to include these deep structures. The inconsistent usage can lead to confusion, and so I have advocated instead the use of the term tectosphere to denote the region occupied by the coherent entities we call plates, retaining for the lithosphere its classical definition as the layer of significant strength. (This usage is not without precedent: in two of the original papers outlining the plate-tectonic theory, Walter M. Elsasser and W. Jason Morgan of Princeton applied the word tectosphere in this context.)

The tectosphere is defined by its kinematic behavior, or purely by its motions; the lithosphere and asthenosphere, on the other hand, are defined by their dynamic behavior, that is, by the

way they respond to imposed forces. Under the oceans the tectosphere and the lithosphere are, for most practical purposes, identical in spatial extent, attaining maximum thicknesses of about 100 kilometers in the oldest ocean basins. Under the continents, however, the tectosphere and the lithosphere are not the same: the cratonic tectosphere extends below the lithosphere, perhaps to depths of 400 kilometers or more.

Continental Thermal Evolution

The variations in tectospheric thickness inferred from the surface-wave and ScS data correlate rather well with estimates of the heat flux coming from below the crust, which suggests that the thickness of the tectosphere is controlled by the temperature structure of the mantle. Tectospheric thickness also correlates with crustal age: generally speaking, the thinnest tectosphere underlies the youngest oceanic crust, whereas the thickest tectosphere underlies the oldest continental crust. In the oceans the plate-tectonic model has provided a simple explanation for these correlations. As new sea floor spreads laterally away from its site of creation at the crest of an oceanic ridge it loses heat by conduction to the surface, and so the temperatures within a surface "thermal boundary layer" decrease with time. Consequently the thickness of the boundary layer increases with the age of the crust, and the heat flux from its surface decreases. Modeling studies indicate that the thickness of this oceanic thermal boundary layer is essentially the same as that of the oceanic tectosphere (and lithosphere); it averages about 70 to 100 kilometers.

The continents also show a systematic decrease in surface heat flow and increase in tectospheric thickness with "crustal age," provided the age is defined not as the time since the original formation of the crust but as the time since the last major orogenic event. This behavior has encouraged some geophysicists to extend the thermal-boundary-layer hypothesis to the continents. According to this hypothesis, the subcontinental mantle is strongly heated during major orogenies and subsequently cools by conduction of heat to the surface. The cratons are therefore explained as regions that have been cooling for a very long time and where the thermal boundary layer has grown to extreme thickness. The times for this thermal decay are actually about right: the thickness of a thermal boundary layer that has cooled for one or two billion years should be nearly 400 kilometers, which could explain the seismic data.

In spite of this agreement and its attraction as a unified theory of tectospheric thermal evolution, there are se-

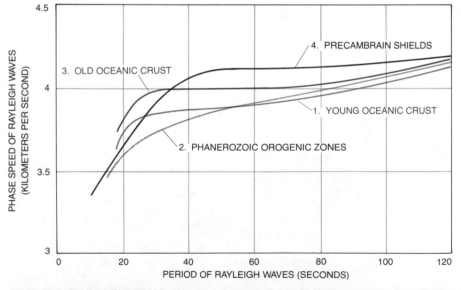

FREQUENCY DEPENDENCE of surface-wave speeds can be used to estimate the variation of the mantle's elasticity with depth. The typical dispersion curves plotted here correspond to regions of crust with different tectonic classifications. In the case of low-frequency surface waves (that is, those with periods longer than about 40 seconds) the highest speeds are observed for waves propagating across the Precambrian continental shields (*curve 4*), whereas the lowest speeds are observed for waves propagating across comparatively young oceanic crust (*curve 1*).

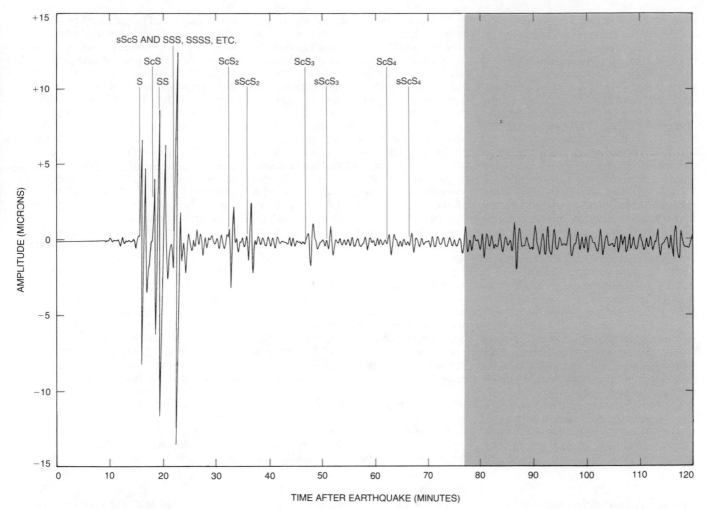

VARIETY OF SHEAR WAVES, seismic waves whose motion is polarized at right angles to the wave paths, are represented in this complex seismographic trace, recorded at a station on the Hawaiian island of Oahu following a deep-focus earthquake near the South Pacific island of Tonga in October, 1974. The various types of transverse waves responsible for the pulses in the seismogram, labeled by the colored symbols at the top, are explained in the illustration on the next page. In general waves that arrived at the station more than about 77 minutes after the earthquake (*colored area*) traveled the long great-circle route around the earth between Tonga and Oahu.

rious difficulties in applying the thermal-boundary-layer hypothesis to the continents. The most obvious problem concerns predictions made by the principle of isostasy. Because no significant shear stresses can be statically supported by the asthenosphere, the pressures exerted on the asthenosphere must be equalized at some minimum depth in the upper mantle called the level of compensation. To a good approximation isostasy requires that the mass contained in columns of equal cross section of ocean, crust and upper mantle be equal; in effect, the sum of the products of the layers' thicknesses and densities must be the same.

Under the oceans as the mantle cools and the thermal boundary layer grows the density within the boundary layer increases. To maintain isostasy the crust and upper mantle subside and the depth from sea level to the ocean floor increases. Hence mantle rock is displaced below the level of compensation and replaced by water, which is much less dense, allowing the mass columns to re-

main in balance. This mechanism explains quite accurately the shapes of the mid-ocean ridges and the location of the centers of sea-floor spreading at the crest of the ridges: the mantle under newly formed crust is hot and light; it cools and subsides as it moves away from the spreading centers.

On the continents, however, the history of vertical movements does not agree nearly so well with the predictions of the thermal-boundary-layer hypothesis. As the continental crust and mantle cool following an orogenic heating event they should also subside, and this subsidence should be accompanied and recorded by the deposit of sediments. (Otherwise all the cratons would be under water!) Although large accumulations of sediments are observed on some continental margins and in certain continental basins, the well-documented history of the cratons is quite different. For example, the growth of a thermal boundary layer 300 kilometers thick—probably the minimum necessary to explain the seismic data for the cratons—

predicts crustal subsidence, and thickening by sedimentation, of nearly 20 kilometers. The cratons, however, are actually characterized by their notable lack of sedimentary cover; huge tracts of basement rock are exposed in the shields, and even the platforms are rarely covered by more than a few kilometers of sediments. Furthermore, cratonic crusts are no thicker than younger continental crusts. It is quite apparent, therefore, that the thermal-boundary-layer hypothesis fails to explain the existence and development of deep continental structure.

What explanation can be offered in its stead? In 1975 I proposed that the thermal evolution of the continental tectosphere is intimately tied to its chemical evolution and that the deep continental structures implied by seismology are the visible expressions of compositional as well as thermal variations in the upper mantle. I was led to this hypothesis by the necessity of constructing a model that meets a particular constraint: a model of a subcontinental tectosphere that is stable over long periods of time in a mantle that must be presumed to be convectively unstable.

Let me explain the reasoning. Below the lithosphere one condition for long-term stability is hydrostatic equilibrium: a rest state in which surfaces of constant pressure and surfaces of constant density coincide and are horizontal. The reason is that in the asthenosphere—below 100 kilometers or so—even small departures from this ideal rest state would cause material to flow. Yet, as I have indicated, the high shear-wave speeds characteristic of deep continental structures seem to require anomalously low temperatures under the continents, and hence lateral thermal gradients in the mantle under the oceans and under the continents, at depths well in excess of 100 kilometers. In a chemically homogeneous mantle such temperature gradients would induce lateral density gradients; surfaces of constant density and surfaces of constant pressure would not coincide. The consequence would be convective instability, and the resulting flow of material would quickly disrupt and eventually destroy the sublithospheric portions of the tectosphere. The deep continental tectosphere is not in fact disrupted and destroyed, however; it is notable instead for its long-term stability.

I proposed, therefore, that the mantle is not chemically homogeneous, and that the lateral thermal gradients associated with continental deep structures are stabilized against convective disruption by differences in composition. The compositional gradients were assumed to be dynamically adjusted in such a way that the tectosphere extending below the lithosphere is in approximate hydrostatic equilibrium with the warmer mantle surrounding it. That is to say, if the mantle were chemically homogeneous, there would be excess mass under the continents owing to the higher densities caused by lower temperatures; according to the hypothesis of inhomogeneous composition, these mass excesses are locally compensated by chemically controlled deficiencies in mass. In other words, the mineralogical assemblages within the thick cratonic root zones are actually equal in density to those at the same level under the oceans because the subcratonic assemblages are cooler; they would be less dense if the continental and oceanic temperatures were the same.

Basalt Depletion

The deduction that continental deep structures are compositionally distinct from their surrounding mantle was based purely on geophysical reasoning, without specific geochemical considerations. Further investigation, however, quickly revealed a simple and geochemically attractive mechanism for generating compositional heterogeneities: the density of the continental tectosphere could be lowered by the removal of a basaltic component from the mantle.

Basalt is the most common magma erupted on the earth's surface: about 20 cubic kilometers of basaltic magma is removed from the mantle every year and added to the crust. Most of this volcanism occurs in the oceans along the ridge-crest spreading centers or at "hot spots" such as Hawaii, but rocks of basaltic composition are also important constituents of the continental crust. When basalt is crystallized near the surface, it is composed primarily of the minerals plagioclase (a calcium-sodium-aluminum silicate) and clinopyroxene (a dark-colored, chemically variable silicate containing iron, magnesium, calcium and sodium). At subcrustal depths, however, basalt crystallizes into a mineral assemblage called eclogite, which also contains clinopyroxene but which has the dense mineral garnet as its aluminous phase instead of plagioclase.

Until 20 years ago some geochemists thought the upper mantle might be composed entirely of eclogite, but now the

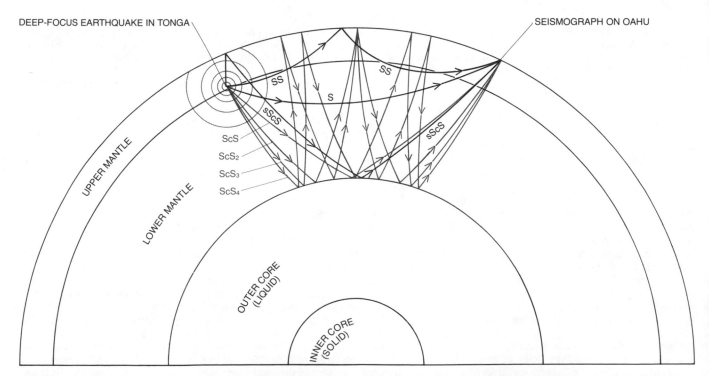

PATHS OF SHEAR WAVES represented in the seismogram on the preceding page are shown in this diagram, along with their conventional seismological designations. The paths labeled S and SS denote transverse waves that travel through the mantle from source to receiver without reaching the earth's core. The waves that travel along the path labeled ScS are reflected once from the sharp discontinuity that separates the solid mantle from the liquid core before they return to the earth's surface. The waves that travel along the paths labeled ScS_2, ScS_3 and so on are reflected more than once from the mantle-core boundary and are also reflected from the surface. The multiple ScS waves are particularly useful for studying the structure of the upper mantle along their paths between source and receiver. Unlike surface waves, shear waves of this type are sensitive to variations in the elasticity of the mantle at great depths. Accordingly they provide critical information about the depth to which certain structural differences between the continental crust and the oceanic crust extend.

bulk of the upper mantle is thought to be a peridotite, which at depths greater than about 70 kilometers forms an assemblage of four minerals: olivine (60 percent), orthopyroxene (12 percent), clinopyroxene (15 percent) and garnet (13 percent). Such a four-phase peridotite is called a garnet lherzolite. Each of the minerals in a garnet lherzolite has a different melting temperature. The basaltic components (clinopyroxene and garnet) melt at lower temperatures than the more refractory components (olivine and orthopyroxene). If the temperature is high enough so that some fraction of the mantle—say 10 or 20 percent—is melted, the molten fraction has a basaltic composition. If this melt is then removed, it leaves behind a residual rock that is depleted in basalt and consists primarily of olivine and orthopyroxene.

The important fact is that at a given temperature and pressure the density of this depleted residual rock is somewhat lower than that of the parental garnet lherzolite; that is, the removal of a basaltic component from the mantle reduces the mantle material's density by a small amount, typically 1 percent or so. The reason for the decrease in density is easy to understand. The removal of basalt leaves the mantle with a smaller proportion of garnet, which is substantially denser than the other minerals in a garnet lherzolite (3.7 grams per cubic centimeter compared with 3.3 grams per cubic centimeter) and also lowers the mantle's content of iron (the heaviest abundant element in the mantle). Surprisingly, these density relations were not generally appreciated by geophysicists and geochemists until quite recently; the first clear, quantitative statements concerning them were published in a short paper by Michael J. O'Hara of the University of Edinburgh in 1975. O'Hara's conclusion that depletion in basalt lowers the density of the mantle has been subsequently confirmed by the experimental work of Francis R. Boyd and R. H. McCallister of the Carnegie Institution of Washington and by my own extensive numerical calculations.

In 1976 I proposed that the compositional variations required to stabilize the continental tectosphere were induced by basaltic depletion of the tectosphere. My calculations showed that the effect should be quantitatively sufficient. At depths of from 150 to 200 kilometers the subcontinental mantle is estimated to be colder than the suboceanic mantle by some 300 to 500 degrees Celsius. In a chemically homogeneous mantle such temperature differences would produce differences in density of from 1 to 1.5 percent. Density differences of that magnitude could be exactly compensated if the peridotite that is now 150 to 200 kilometers below the

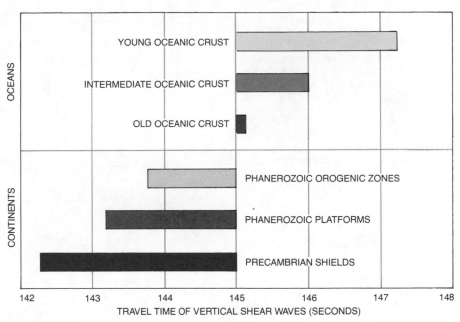

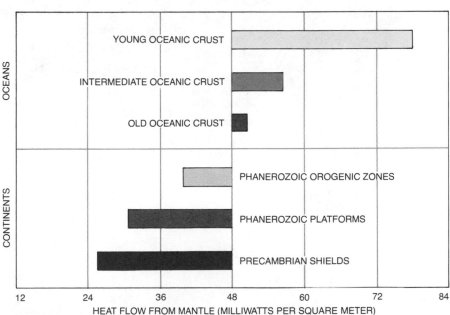

CLOSE CORRELATION is observed for different types of crustal rock between the time it takes for a shear wave to travel vertically from a depth of 700 kilometers to the surface (*bar charts at top*) and the amount of heat flowing from the mantle (*bar charts at bottom*). The vertical travel times are averages constructed from a large number of multiple *ScS*-wave travel times; the mean travel time for the entire globe is about 145 seconds. The characteristic heat-flow values shown are also averages based on the measured heat flow for the various types of crust; the mean global heat flow is 48 milliwatts per square meter. The correlation supports the view that the seismic differences associated with different types of crust arise from differences in the thermal structure of the underlying mantle. Colors are keyed to map on pages 4 and 5.

cratons had been melted to the extent of 10 to 20 percent at some time in its history, and if the basaltic molten fraction had been removed by migration to the crust.

This basalt-depletion hypothesis can be checked by direct geochemical observations. Rocks that were once 150 to 200 kilometers below the continents are actually found on the surface, as rounded pebbles and boulders called xenoliths (foreign rocks) in formations known as kimberlite pipes: the eroded necks of

certain peculiar volcanoes found only in stable continental regions [see "Kimberlite Pipes," by Keith G. Cox; SCIENTIFIC AMERICAN Offprint 931]. These pipes have been studied extensively because they are the ultimate surface sources of all diamonds. (The type locality of kimberlite—the peridotitic volcanic rock that fills the pipe—is the famous diamond-producing area of Kimberley in South Africa.) The xenoliths, some containing diamonds, apparently were ripped from the walls of the volcanic

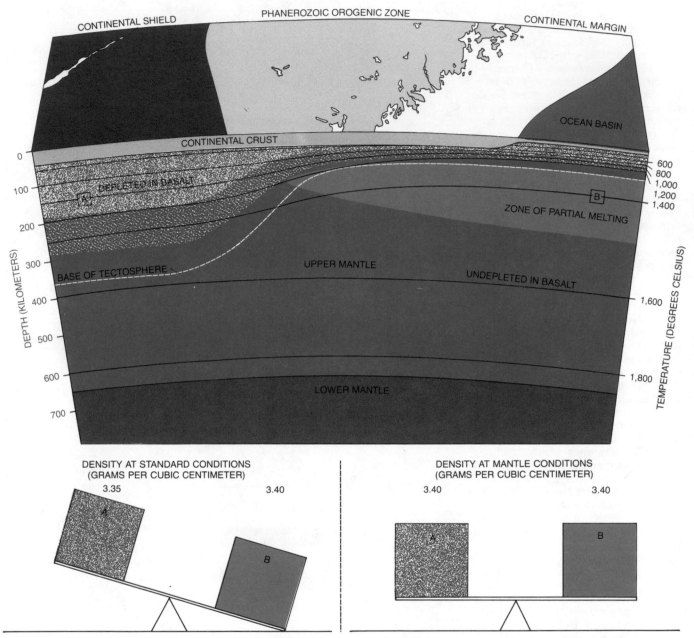

CONTINENTAL SHIELD   PHANEROZOIC OROGENIC ZONE   CONTINENTAL MARGIN

OCEAN BASIN

CONTINENTAL CRUST

DEPLETED IN BASALT

ZONE OF PARTIAL MELTING

BASE OF TECTOSPHERE

UPPER MANTLE

UNDEPLETED IN BASALT

DEPTH (KILOMETERS)

LOWER MANTLE

TEMPERATURE (DEGREES CELSIUS)

DENSITY AT STANDARD CONDITIONS
(GRAMS PER CUBIC CENTIMETER)

3.35   3.40

DENSITY AT MANTLE CONDITIONS
(GRAMS PER CUBIC CENTIMETER)

3.40   3.40

DENSITIES OF MANTLE ROCKS at a given level under the continents and the oceans are assumed to be nearly equal in the author's model, even though the temperatures under the continents are lower. To compensate for an increase in density due to thermal contraction he postulates that the mantle rocks are differentially depleted in certain basaltic components, which under the conditions present in the mantle are represented by the minerals clinopyroxene and garnet. At standard conditions the peridotites that constitute the cold continental root zones would accordingly be less dense than their suboceanic counterparts. At a depth of 150 kilometers, for example, the subcon- tinental mantle (*A*) is estimated to be about 400 degrees Celsius cooler than the suboceanic mantle (*B*), requiring a decrease in density by the basalt-depletion mechanism of approximately .05 gram per cubic centimeter (*shown in the scales at bottom*). Standard conditions here mean a temperature of 25 degrees C. and a pressure of one atmosphere; mantle conditions are temperatures of 1,000 and 1,400 degrees respectively for subcontinental and suboceanic mantle rocks and a pressure of 50,000 atmospheres for both. Diagram at the top is identical with the one on page 3, except for the substitution of the temperature contours (*black*) for the seismic-velocity contours.

conduit during the rapid ascent of gaseous kimberlite magma from a great depth and were transported quickly to the crust, in most cases without significant chemical alteration. Geochemical and petrological studies have shown that many of the xenoliths are samples from depths of between 100 and 250 kilometers.

Most of the mantle xenoliths are garnet lherzolites. Petrologists such as P. H. Nixon of the University of Papua New Guinea and Boyd of the Carnegie Insti- tution have noted that the xenoliths are apparently depleted in a basaltic component: they typically contain less garnet and clinopyroxene than does the mineral assemblage estimated to constitute the bulk of the oceanic upper mantle. To demonstrate the depletion I constructed a model composition of the cratonic upper mantle in the depth range from 150 to 200 kilometers by averaging together the available chemical analyses of garnet lherzolite xenoliths from kimberlite pipes. The depleted na- ture of this average continental garnet lherzolite is shown in the illustration on page 106. At a standard temperature and pressure the density of this continental material turns out to be 1.3 percent less than the density of the best available model for the oceanic upper mantle—or right in the middle of the 1-to-1.5-percent range specified by the basalt-depletion hypothesis!

Closer examination of the compositions and densities of the garnet lherzolite xenoliths provides a more stringent test of the basalt-depletion hypothesis. The differences in composition between the continental and the oceanic mantle are hypothesized to be proportional to their temperature differences; the temperature differences can be presumed to decrease with increasing depth, vanishing near the base of the tectosphere. The amount by which the continental rocks are depleted in basalt should therefore decrease with depth. As was first pointed out by Nixon and Boyd, that is indeed the case. The density contrast appears to decrease with depth just as predicted by the basalt-depletion hypothesis.

The basalt-depletion hypothesis was motivated indirectly by seismological data: the high wave speeds that indicate low temperatures under the cratons. Basalt depletion may play a more direct role in explaining those seismic-wave speeds, and particularly the shear-wave speed. Two effects are important. First, basalt removal lowers the iron content of the mantle, and experiments have shown that seismic-wave speed rises with decreasing iron content. Second, basalt removal increases the melting temperature of the mantle (because the components that melt at low temperatures are withdrawn) and that also increases the wave speeds, even at temperatures below the melting point. Basalt depletion thus helps to explain why the shear-wave speeds under the cratons are so high.

In summary, the basalt-depletion hypothesis has several attractive features: it provides a mechanism for stabilizing the sublithospheric portions of the continental tectosphere against convective disruption, it explains at least two salient features of kimberlite xenolith petrology and it is consistent with the available seismological constraints. Much more extensive testing is nonetheless needed before the model can be accepted as a foundation on which to construct a theory of continental development. Diagnostic tests of the model do not come easily; the phenomena it represents are buried under hundreds of kilometers of rock. Yet the constraints on the nature of the deep continental root zones are multiplying at such a rapid rate that if the model is fundamentally wrong, we should know it very soon.

Cratonic Stability Reassessed

In the meantime some tentative speculations on the nature of cratonic stability and continental evolution are warranted. In the context of the model I have proposed, continental drift is dynamically complex, more than simply a passive rafting of light crustal material on an otherwise undistinguished piece of lithosphere. The basalt-depletion model suggests that continental tectonic behavior is actually regulated by chemi-

cal variations extending deep into the upper mantle.

Under regions of active tectonism at or near plate boundaries, such as the western U.S. and southern Europe, the layer of depleted peridotite is apparently thin. As the mantle under this thin layer cools beyond some critical point, undepleted mantle material sinks, causing convective overturn and the transport of more heat to the surface. Because they are heated by this convective overturn, the crust and the uppermost mantle in such a region are weak and easily deformed. In contrast, the depleted layer under the cratons is hypothesized to be much thicker. The low density of this residual peridotite prevents it from being easily mixed back into the mantle by convective action, so that the material in the depleted zone is stabilized in nearly hydrostatic equilibrium at lower temperatures than its surrounding mantle. Because its temperature is lower and its low-temperature melting fraction has been removed its viscosity is higher, and so the zone of horizontal shearing associated with plate motions is confined below the depleted zone. The

crust and uppermost mantle are isolated from the thermal perturbations and mass motions associated with small-scale convective disturbances, and so cratonic stability is maintained.

How is a thick, cool, depleted subcrustal tectosphere formed in the first place, however? Where does all that basalt come from and where does it go? Geologists have deduced that the basic plate-tectonic processes observed today—sea-floor spreading, continental drift, subduction and so on—were active in the Precambrian era, which ended about 600 million years ago. Some geologists believe there is evidence for plate tectonics in rock formations three billion years old or more. It is reasonable to assume, therefore, that tectospheric development has been governed by the plate-tectonic cycle, at least for the past three billion years.

What particular processes may have been important in depletion? At present the continental upper mantle is being depleted in basalt in two primary settings: at sites of the magmatic episodes associated with continental rifting and in the wedge of mantle above the sub-

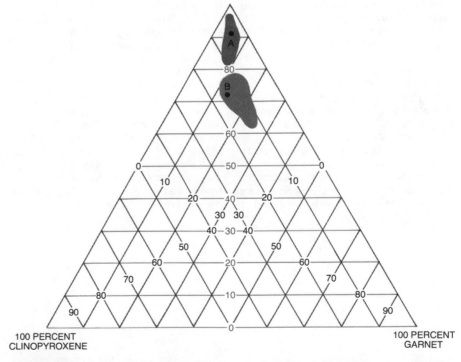

100 PERCENT OLIVINE AND ORTHOPYROXENE

100 PERCENT CLINOPYROXENE

100 PERCENT GARNET

EVIDENCE supporting the basalt-depletion hypothesis has been obtained from xenoliths (literally "foreign rocks") brought to the surface during the eruption of basaltic magma from kimberlite pipes. These rocks, which presumably are representative samples of the composition of the upper mantle under the continental crust at depths ranging from 100 to 250 kilometers, are found to have smaller concentrations of clinopyroxene and garnet (*upper shaded area*) than those estimated for the upper mantle under the oceanic crust (*lower shaded area*). The density of the kimberlite xenoliths is on the average about .05 gram per cubic centimeter less than the estimated density of the peridotite in the oceanic upper mantle, in agreement with the value predicted by the basalt-depletion hypothesis for the average density of the peridotite under the continental crust. The composition of the basalt-depleted peridotite at point A in this illustration and in the one on page 12 is 67 percent olivine, 23 percent orthopyroxene, 4 percent clinopyroxene and 6 percent garnet; the undepleted peridotite at point B in both illustrations is 60 percent olivine, 12 percent orthopyroxene, 15 percent clinopyroxene and 13 percent garnet. The basaltic components are shown in black; nonbasaltic components are shown in color.

duction zones. The former are characterized by huge volumes of "flood basalts" extruded during periods of crustal extension preceding the breakup and drifting of continents. (For example, the Paraná basin flood basalts of South America, which were erupted just before the opening of the South Atlantic Ocean, have a total volume of nearly two million cubic kilometers; the volume of mantle depleted by this episode must have been at least several times greater.)

Even more voluminous are the basalts emplaced in the crust above subduction zones, landward of the oceanic trenches along the "active" continental margins and island arcs. These sites are usually associated with the more silica-rich andesitic volcanism, but basalt is actually the more common magma type, particularly in juvenile island arcs. Petrologists and geochemists are trying to identify the sources of these magmas. The andesites may come from the subducted oceanic crust, but most of the basalt is apparently generated from the mantle wedge above the descending oceanic lithosphere. Hence a major portion of the residual depleted mantle eventually incorporated into thick tectospheric root zones may be originally depleted in regions of rapid island-arc formation such as the present-day western Pacific.

Melting and basalt extraction are processes confined mainly to the mantle above a depth of 200 kilometers, however; below that the pressure is so great that the melting required to produce basalt is rarely achieved, if ever. A depleted tectosphere thicker than 200 kilometers therefore cannot be generated by melting processes alone. E. R. Oxburgh and E. M. Parmentier of the University of Oxford have recently proposed another source. They suggest that some of the peridotite that is depleted as it rises and melts at a mid-ocean ridge and is transported by sea-floor spreading to a site of subduction may in the course of subduction attach itself to the tectosphere.

However it is generated, any proto-continental tectosphere in an island-arc environment or along an active continental margin would be thin, chemically heterogeneous and poorly consolidated. The thickening comes with time. First the dispersed regions of depleted mantle and their overlying crust are swept up and accreted by plate motion to the primary continental masses. Although such terrain is somewhat consolidated, it still has a thin tectosphere and hence may be a site of orogenic activity; good examples of continental tectosphere in this stage of development are found today in Indonesia, western Canada and Patagonia.

Further consolidation and thickening of the tectosphere apparently result from the major compressive events at convergent plate boundaries, particularly episodes of collision between continents. The history of the crust following a continental collision has been analyzed in detail by John F. Dewey and Kevin C. A. Burke of the State University of New York at Albany. During very violent collisions, such as the ongoing one between India and Asia, the former continental-margin crust is compressively thickened by a factor of two or more, causing the uplift of a plateau (such as the Tibetan plateau, where surface-wave data indicate that the crustal thickness is about 70 kilometers). Dewey and Burke think this process of compressive crustal thickening involves deformation over a broad zone. Such deformation should be concomitant with a proportional thickening of the subcrustal tectosphere by the lateral and downward advection of depleted peridotite. In this way the thick root zones of the cratons could be formed.

Depletion, consolidation and thickening, then, are thought to be the important constructive events in the development of a continental tectosphere. To stabilize a thick tectosphere in any particular region, these processes may have to be repeated over and over again because they must often compete against destructive events caused by fluid-mechanical instabilities in the upper mantle. Whereas large-scale basaltic depletion is nearly irreversible, consolidation and thickening of the tectosphere surely are not. Any significant heating of a cool, thick, stabilized tectosphere by heat conduction from its periphery, by intrusion or by internal heat sources results in its thinning and dispersal by the upward and lateral flow of low-density depleted mantle. Such motions may play an important role in episodes of continental rifting and breakup. A heating event of this kind may have destabilized the once cratonic portions of the western U.S., causing the widespread extension and rifting now taking place in the Basin and Range province of Utah, Nevada and surrounding states.

The rough sketch of continental tectospheric structure and development presented in this article is no more than an incomplete and highly speculative working model, shaped by many disputable generalizations and colored by the author's own prejudices. Much closer scrutiny is necessary to evaluate its feasibility and to validate the hypotheses on which it is based. Regardless of its eventual fate, however, the model does emphasize the maturity of modern earth science. Any detailed picture of deep continental structure and its evolution must be painted on a cloth woven from many threads of geological, geochemical and geophysical information. Only such a unified picture will reveal the true nature of the dynamic processes that have shaped the continents.

VOLCANOES AND HEAT FLOW

III

INTRODUCTION

Only volcanic eruptions can match earthquakes as dynamic disruptions of the earth's surface. The four articles included in this section describe the grandeur, the complexity, and the widespread distribution of volcanic activity. They also discuss the underlying causes of the majestic volcanic mountain chains, island arcs, and basalt plateaus that garland the globe. Since ancient times, the very existence of many types of eruptions, from nuées ardentes (glowing avalanches) and viscous lava flows to geothermal areas with geysers and hot springs, has been a puzzle. Nowadays, as these articles indicate, new mechanical and geochemical models for volcanism are emerging.

Recent analysis depends heavily on the model of the continuous creation, horizontal drifting, and subduction of tectonic plates. But, as yet, no one evolutionary explanation for volcanism within continents, in the mid-oceans, and along the island arcs is available. We are at a fascinating period of ingenious but competing theories supported by largely circumstantial evidence from structural geology, petrology, geophysics, and seismology.

Because the formation of molten rock and the rise of magma to the earth's surface involves heat exchanges in the interior, the search for causes of volcanism must be based on a knowledge of the earth's interior temperature patterns. Volcanic eruptions and large earthquakes represent spectacular releases of energy narrowly confined in space and time, but there is a slow, persistent loss of energy from the interior continually taking place everywhere over the earth. This energy is the heat being conducted to the earth's surface through the outer shell, or lithosphere. If we accept that the internal heat is the energy source for the dynamic behavior of the earth—of which mountain building, drifting plates, earthquakes, and volcanoes are surface manifestations—then the conductive heat loss at the surface represents the exhaust of this great heat engine.

The last two articles in this section present complementary aspects of the convection model for the earth's driving forces. This is now by far the most popular theory to explain the movement of tectonic plates, but, as the articles indicate, buoyant overturn of plastic mantle rock must involve complicated physical processes. Even the geometrical patterns of flow are still speculations, with some geophysicists favoring convection currents through the whole mantle and others favoring smaller, more superficial currents. Once again, earthquakes and volcanoes play a part in these intriguing structural speculations.

Q: What is a volcano?
A: A volcano is a place where molten rock and gas (magma) emerge onto the surface of the earth. The mountain or hill formed by the eruptive debris is also

often called a *volcano.* Over 500 volcanoes are known to have erupted during the past 500 years, killing about 200,000 people.

Q: Where do volcanoes occur?

A: Eruptions of magma occur mainly along the mid-oceanic ridges and along the subducting margins of tectonic plates. The greatest number of unsubmerged volcanoes are steep-sided mountains of lava and ash, forming island arcs around the Pacific Ocean.

Q: Do all regions of mountain building allow magma to escape to the surface?

A: No. Not all young mountain ranges have volcanoes. For example, the Himalayas and the Alps, which formed by collisions of continental plates, do not have volcanic activity.

Q: What is the source of magma, and how is it produced?

A: Some regions of the earth's outer shells at depths of 100 km or less have temperatures, pressures, and compositions that produce melting of the rocks. This magma, which is lighter than the overlying rocks, rises through buoyancy. At mid-oceanic ridges, tholeiitic basalt differentiates from the mantle rock (probably peridotite) at a depth of about 25 km and at a temperature of about 1,250° C. The rising magma erupts as pillow lavas and intrudes as dikes into the sea floor, forming a thin (5 km) oceanic crustal layer. The sea floor moves away from the ridges, and the material of the crust cools. At the deep ocean trenches, the plate moves downward into the mantle as a cool slab of lithosphere; as it comes in contact with hotter rock, its temperature begins to increase. Strains build in the subducting slab as it is bent and heated, thus producing earthquakes. It is suggested that the partial melting of the subducting oceanic crust and its veneer of wet sediments produces the typical andesitic basalt magmas seen in island arc volcanoes. In this model, the magma at subduction margins originates from depths of 50 to 100 km beneath the surface. Other types of volcanoes, both continental and oceanic, occur at isolated locations within the plates.

Q: What are the main types of volcanoes?

A: There are many types of volcanoes, both in shape and in form of activity. *Shield* volcanoes, such as those of Hawaii, form gentle slopes from thousands of thin flows of liquid lava. At the other extreme, such steep-sided edifaces as Lassen Peak in California are produced by very viscous lava, which is thrust up through a feeding pipe to heights of hundreds of meters. Also, there are *composite* volcanoes with concave slopes that steepen to the summit, such as Mt. Rainier in the United States and Mt. Mayon in the Philippines. These consist of alternating layers of volcanic ash and lava flows. A different type of volcano is produced when the discharge of magma occurs not through a central conduit but along fissures in the crust. The outpouring of lava and ash then produces an elevated *plateau,* like the basaltic plateaus of the Pacific Northwest in the United States. These were formed between 10 and 20 million years ago by successive lava flows ultimately more than 2,000 meters thick and 350,000 square km in extent.

Q: What forms the caldera at the top of many volcanoes?

A: Contrary to widespread belief, the huge basins, sometimes several kilometers across, at the tops of many volcanoes are produced not by explosive discharge of volcanic gases and bombs but by collapse of the top of the volcano. Cataclysmic venting of gas and pumice empty magma reservoirs below the summit, and the mountain tops fall into the unsupported depths. Crater Lake in Oregon is an example of a volcanic caldera.

Q: How much energy is released in a volcanic explosion?

A: A single catastrophic explosive eruption, such as Krakatoa in 1883, might release about 10^{25} ergs, or a few times more than the energy of a magnitude 8.5 earthquake. A single eruption may produce several cubic kilometers of lava and ash. Such large eruptions are, however, infrequent. The total erupted volume of lava on the sea floor has been estimated to be about 4 cubic km per year, and, on land, less than 1 cubic km per year. The total amount of heat energy brought to the surface by all volcanic activity each year is several orders of magnitude less than that conducted out through the whole earth's surface by nonvolcanic processes.

Q: Do active volcanoes occur along plate boundaries apart from the subduction zones or mid-oceanic rises? Do they occur within plates?

A: There are few volcanoes along lateral boundaries where plates slide by each other, but a number of magnificent ones lie within the plates. For example, the southeast end of the Hawaiian archipelego is thought to mark a "hot spot" that is fixed in the mantle beneath the lithosphere. As the Pacific plate moves northwest, the volcanic islands are carried away from this hot spot, and another volcano is formed, producing a chain of volcanoes of sequential age.

Q: What is the most important factor in producing eruptions?

A: There is little doubt that the gas content of the magma is the key factor. Viscosity of magma is a function of the chemical composition, temperature, and the amount of dissolved gas. At the volcanic vent, over 90 percent of the gas emitted is steam, with lesser amounts of carbon dioxide, sulfur, and other vapors. The sudden release of gas dissolved in the magma produces effervescence and heat-yielding reactions, which lead to fountains of boiling lava and gaseous releases.

Q: Can volcanic eruptions be predicted?

A: The geographical location of likely eruptions is known rather well, because volcanic activity is largely confined to regions of recent mountain building and the fractured margins of plates. Likely geographical locations are usually associated with significant earthquake activity and are marked by recent volcanism. The time of an eruption is quite difficult to forecast, but the use of such geophysical devices as seismographs and tilt meters around the volcano has proved helpful.

Q: How can volcanic risk be controlled?

A: There is a risk of damage on the flanks of active volcanoes, but no single eruption is likely to affect seriously more than a small proportion of the total area. Preparation of topographic maps can identify likely courses of lava flow, and wind pattern maps help predict the distribution of ash falls and toxic gases. Sometimes dams and channels can be built to divert lava and volcanic mud flows.

Q: Can the geothermal energy associated with volcanism be harnessed?

A: The thermal energy in volcanic areas is increasingly being tapped. Electric power from steam vents has been used commercially, for example, in Italy, New Zealand, Mexico, and California. At the Geysers, California, the magma chamber is at a depth of about 8 km. Dry steam is tapped in wells drilled into the faulted crustal rock and fed to turbine generators, which in 1979 produced over 500 megawatts of electrical power, enough to meet the needs of a city the size of San Francisco.

Q: How much heat is the earth losing as a result of conduction from the interior to the surface?

A: Heat is continually being transferred to the oceans and atmosphere from the earth's interior. The average amount of heat lost in this way is about 30 trillion watts over the entire planetary surface. By comparison, the amount of energy arriving from the sun is about 6,000 times greater.

Q: What are the sources of the earth's internal heat?

A: There is a store of heat left over from the early molten condition of the earth, and most common rocks contain radioactive elements whose decay also liberates heat. Tidal friction from slowing of the earth's rotation may also make a contribution to the earth's internal heat.

Q: What is the amount of heat lost per unit area by conduction through the earth's surface? Does it vary much from place to place?

A: Most observations of heat flow range from 20 to 180 milliwatts per square meter, with a global average of about 60 milliwatts. Although patterns of heat flow are different in continental and oceanic regions, the most frequently *observed* values (the modal averages) are the same for oceans and continents—the actual heat flow in young oceanic regions may be masked by water circulation in crustal rocks.

Q: How are heat flow measurements made?

A: Heat flow is not measured directly but depends on the temperature gradient and thermal conductivity of the rocks. Under the oceans, a long, heavy, hollow cylinder is dropped into the soft sediments, and the temperature is measured at intervals along the cylinder with fixed electric thermometers. In continents, boreholes are drilled into the rock (or mines are used), and thermometers are placed at various levels. The thermal conductivity of the rock is also required, and this is measured from samples in the laboratory.

Q: What does the geographical distribution of the global heat flow look like?

A: There are now more than 5,000 heat flow measurements from around the world, although the distribution is still uneven geographically. Heat flow is usually high in volcanic and geothermal areas. It decreases with the increasing age of the ocean floor, and there is also a less marked decrease in heat flow with increasing age of geological provinces in continental areas.

Q: What is the heat flow pattern across island arcs?

A: The general pattern across subduction zones, such as the Japanese island arc, has low heat flow near the ocean trench and high heat flow on the landward side of the arc. The low heat flow zone suggests that, as the relatively cool slab plunges downward, it absorbs heat from the earth's mantle. The volcanoes of the arc and the region of high heat flow beyond them suggest frictional heating of the rocks at the top side of the subducting slab.

Q: Can an entirely observational heat flow map now be constructed for the globe?

A: Heat flow measurements are still not sufficiently well distributed for an analysis to be based on observations alone. However, extensive extrapolation into unsurveyed areas is possible, on the assumption that heat flow is a function of the age of the crust. A heat flow map of the globe constructed in this way shows significant world-wide variations in heat flow. Oceanic ridges can be seen as heat flow highs, and the Pre-Cambrian continental shields as heat flow lows.

Q: What are the main inferences that can be drawn from the world heat flow map?

A: Seismic waves provide the traditional method for estimating interior earth structure and crustal thickness. However, inferences on thicknesses of the crust and lithosphere can also be drawn from the global measurements of heat flow. The depth at which the inferred temperature-versus-depth curve intersects the calculated melting curve of the mantle rocks determines the thickness of the lithosphere at that place. Thus, heat flow data provide some idea of the variations in thickness of the lithosphere. Results indicate that the lithosphere is thinnest along oceanic ridges, where the plate thickness is only a few kilometers. The lithosphere may increase to over 100 km under the oldest ocean basins and several hundred kilometers under the old continental shields.

Q: What contributes to the heating of the lithospheric slab in subduction zones?

A: First, heat conducts into the descending cooler slab from the surrounding warmer mantle. Second, heat is presumably generated by friction and viscous dissipation at the boundaries between the slabs and the surrounding mantle. Third, the radioactive elements in the earth's crust add heat at a constant rate to the descending slab. Finally, mineralogical changes taking place as the rocks in the plate are subjected to increasing pressure may release heat energy.

Q: What is the rate of the descent of subducting slabs?

A: Estimates have been made for over fifteen cases of subduction. The velocity varies from 0.5 cm per year for the South American plate under the Caribbean to 9.5 cm per year for the Cocos plate under the North American plate. The average subduction rate is approximately 8 cm per year.

Q: What is the variation of gravity across subduction zones?

A: The overall effect is the sum of anomalies from various structural changes. The oceanic trench produces a gravity low, whereas the cooler and denser lithospheric slab gives rise to a positive gravity anomaly. Variations in the thickness of the crust from oceans to island arcs to continents add more complications. Models of subducting plates, such as that under the Andes, yield theoretical gravity anomalies that are close to those observed. Uncertainties in the density variations, however, make the models nonunique.

Q: Why are earthquakes regarded as the most direct evidence for subduction slabs?

A: Intermediate and deep focus earthquakes occur along a zone (named after H. Benioff) that dips under the island arc from the ocean trench at an angle that is commonly about 45 degrees. However, some Benioff zones have shallow dips, whereas others have nearly vertical dips. The deepest earthquakes have foci at 650 km, approximately the depth at which the slab would pass from brittle to plastic behavior. In island arcs such as the Aleutians and Japan, where detailed studies of focal depths and earthquake mechanisms have been made, earthquake foci lie in a narrow zone some 20 to 30 km thick near the top of the slab. Very recent studies show that, at some places, there are two parallel bands of foci, with the deeper band toward the cooler center of the subducting lithosphere. The focal mechanisms of the earthquakes indicate stresses in the slab which are consistent with a bending, brittle plate under gravitational forces.

Q: What are the elastic properties of the subduction zone?

A: Seismic velocities in the descending slab are from 10 to 15 percent higher than those in the asthenosphere on each side. The attenuation of high frequency seismic waves that pass up through the slab is less than that of high frequency waves that pass outside the slab.

Q: What is the postulated scale of a convective flow in the earth's mantle?

A: The motion of a plate from a mid-oceanic ridge to a subduction zone may be conceived of as the top half of a large-scale convection loop. For the Pacific plate, this large-scale circulation has a horizontal dimension of 10,000 km.

Q: What is the most significant parameter for mathematical description of fluid convection?

A: For the mantle of the earth, this is the *Rayleigh number*, which is proportional to the ratio of the time needed to heat a fluid layer by conduction and the time needed for fluid particles to circulate once around the cell. Quite large Rayleigh numbers occur in the mantle, with a lower limit of about 10^6. For small Rayleigh numbers, thermal convection cannot occur; for large Rayleigh numbers, the system becomes unstable, and disturbances grow into convection cells. Some laboratory experiments with convecting oil in two-dimensional configurations have given some idea of the complexity of patterns that can arise when the fluid is driven by heating from below. Such analogies cannot be carried too far, because the geographical pattern of tectonic plates suggests three-dimensional cells of various sizes within the earth. Also, the rocks of the mantle are not only heated from below but contain heat sources within them, and they have viscous properties that change with temperature and pressure.

Q: What are the main models for convection in the earth's mantle?

A: One vigorously supported model assumes that mantle convection extends from the top of the mantle to the bottom at a depth of 2,900 km. In the second commonly discussed model, the large-scale convection is confined to the outer 650 km, which is the depth of the deepest earthquakes. (This model is not inconsistent with the existence of much larger convection cells below 650 km.)

Q: Is there observational evidence that can be used to distinguish one convection hypothesis from another?

A: Unfortunately, tectonic plates are rigid enough to mask most of the effects that might be associated with small-scale convection of cells of a few hundred kilometers in diameter. Regional variations in gravity and in the depth of the ocean, however, may provide information on such small-scale convection. Such cells under the oceans also provide one explanation for the chains of volcanoes, such as the Hawaiian arc, for which the magma source beneath the lithosphere does not move as fast as the oceanic plate.

Volcanoes

by Howel Williams
November 1951

The awe-inspiring mountains of melted rock reflect the dynamic activity of the earth's crust. Although they have caused much death and destruction, they are also useful and could be more so

DURING the past 400 years some 500 volcanoes have erupted from the depths of our planet. They have killed 190,000 people; the most destructive eruption, that of Tamboro in the East Indies in 1815, wiped out 56,000 in one gigantic explosion. Volcanoes understandably have always terrified mankind. Yet it should not be forgotten that they also play a constructive role for our benefit. It is not merely that volcanic eruptions have provided some of the world's richest soils—and some of our most magnificent scenery. Throughout geologic time volcanoes and their attendant hot springs and gas vents have been supplying the oceans with water and the atmosphere with carbon dioxide. But for these emanations there would be no plant life on earth, and therefore no animal life. In very truth, but for them we would not be here!

What exactly are volcanoes, and how are they formed? Obviously they are symptoms of some kind of internal disorder in the earth. The eruptions we see at the surface are only small manifestations of great events going on below, events about which we can only speculate. We do, however, have some clues to what may be happening—a few tantalizing points of light that make volcanoes a most fascinating field of study.

The first clue lies in the location of the volcanic regions on the world map. We know that the active volcanoes are concentrated in parts of the world where earthquakes are most common, particularly in those areas where the earthquakes have a tendency to originate at a level about 60 miles down in the earth's crust. This suggests that volcanoes are connected with disturbances in the earth at that depth. Secondly, we know that most of the world's volcanoes are in young mountain belts, that is, where the face of the earth has recently been wrinkled and cracked.

Tens of miles below the surface of the earth there is an extremely hot shell of glassy or crystalline material. This solid

SMALL VOLCANO ERUPTS from the caldera of the larger Okmok volcano on Umnak Island in the Aleutians. A caldera is a large crater that is formed when a volcano ejects so much material that its cone collapses.

material becomes liquefied if the pressure on it is reduced or the temperature rises. The pressure may be reduced by the bending or cracking of the rocks lying above it; the temperature may be increased by radioactive heating. In either case, the liquefied material forms a fluid mass, called magma, that is lighter than the overlying rocks, and it tends to rise wherever it finds an opening. If there are fractures in the rock that let it rise directly to the surface of the earth, it comes out quietly as a flood of fluid lava. Sometimes it reaches a roof of solid rock a few miles below the earth surface; in that case it may spread sideways and form a reservoir of hot magma that erupts into one or more vigorous volcanoes where it finds cracks in the roof.

Disturbances of the earth in regions of mountain-building produce conditions favorable to formation of molten magma and its escape to the surface. To be sure, not all young mountains have volcanoes; there is none in the Alps or the Himalayas. These mountains have an unusual structure that suppresses eruptions. They were formed by low-angle thrusting and overfolding of the earth's skin; one layer is piled on another, making a thick cover of rock through which magma does not escape. In the mountain belts where volcanoes do occur there is less overlapping of the rock layers; these mountains have steep fractures that go deep into the earth.

The geological record shows that volcanic activity is most widespread during the period of adjustment that follows the formation of mountain ranges. After the great spasms of folding and uplifting that create the mountains have subsided, the earth's crust tends to settle down to a condition of stability, and it is during this stage of adjustment that the eruption of continental volcanoes reaches a maximum.

Types of Volcanoes

A volcano is usually pictured as a cone with a crater at the top which from time to time blasts forth streams and glowing bombs of lava and shattered rock. Actually there are almost as many types of volcanoes as there are landscapes. They range from the explosive kind to the sluggish and gentle, and they come in a great variety of shapes and sizes. The form a volcano takes depends not only on the structure of the earth below it but also on the physical nature of the erupting magma, or lava. One of the most important factors determining the shape and activity of a volcano is the magma's viscosity. This varies greatly; some lavas are so fluid that they flow over the ground at more than 20 miles an hour; others are so viscous that they move at little more than a snail's pace, and even

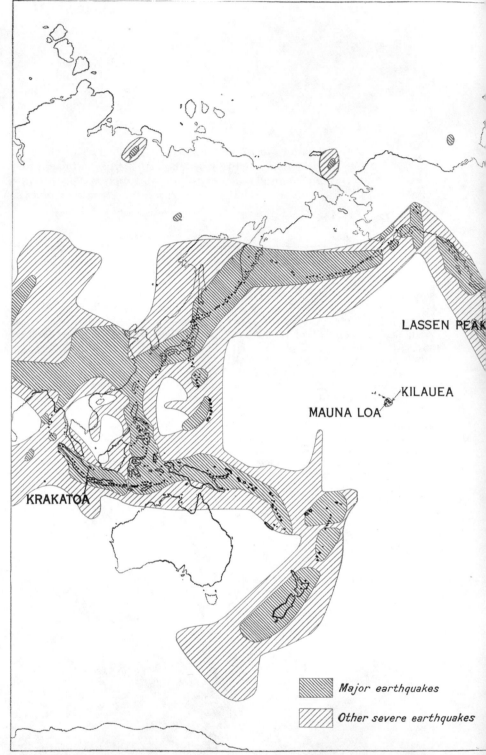

VOLCANOES AND EARTHQUAKES tend to occur in the same parts of the world. On this map the outlines of the land are shown in gray. The regions in

the strong blow of a pick scarcely dents their incandescent surfaces.

Usually the more fluid the magma, the more extensive is the flow of lava, the flatter the resultant edifice and the fewer and weaker the explosive eruptions. A volcano formed mainly by quiet effusions of liquid lava generally has the shape of an inverted saucer. The vol-

canoes of Hawaii are of this kind, and they illustrate various stages in its growth. During the early stages of formation of such a volcano copious streams of extremely hot and fluid basalt are discharged from two or three intersecting rifts in the rock at the earth's surface. Where the rifts intersect a small summit-crater forms. As the volcano

POPOCATEPETL

MT PELÉE

KRAKATOA

which major or severe earthquakes have taken place are depicted in red. The volcanoes are shown as black dots. Six discussed in the text are labeled. The map was prepared by the American Geographical Society.

grows to maturity, the summit crater is much enlarged by gradual collapse of its surrounding walls, and lines of pit-craters develop along the rift-zones cutting the flanks of the volcano. The Hawaiian volcanoes Kilauea and Mauna Loa are now in this stage of evolution. Later, in the volcano's old age, new lava flows fill up and obliterate the summit-and pit-craters. Eruptions take place at longer intervals; the lavas become more varied in composition, and, because most of them are more viscous than the earlier flows and therefore stick on the sides of the mountain near the top, the upper part of the volcano becomes increasingly steep. At the same time, because the longer intervals of rest permit development of greater gas-pressure in the viscous magma, explosive activity becomes more frequent and violent. Cones of ash grow in clusters on the higher flanks of the mountain. Mauna Kea and Kohala on Hawaii are now in this stage of old age, and Hualalai has lately entered it.

At the opposite extreme there are vol-

MAUNA LOA in Hawaii is shaped like an inverted saucer. This type of volcano results from successive out- pourings of relatively liquid lava. In foreground are cin- der cones on the flank of the older volcano Mauna Kea.

canoes formed by lava squeezed out of the earth in an exceedingly viscous con- dition, somewhat like toothpaste from a tube. This produces a very steep-sided mountain. Indeed, the lava may be so nearly solid when it is thrust up through its "feeding pipe" in the earth that it rises as a slender obelisk, like the one that was pushed up to a height of 1,000 feet on top of the dome of the celebrated Mount Pelée in the West Indies in 1902. Lassen Peak in California is another good ex- ample of a viscous protrusion.

Other volcanoes, such as Mount Shas- ta and Mount Rainier in this country, Mount Mayon in the Philippines, Oriza- ba and Popocatepetl in Mexico and Fuji- yama in Japan, are built in part by out- pouring of lava and in part by the explosive discharge of fragments of rock. These so-called composite volcanoes have concave slopes that steepen to the summit. Their graceful profile rises from a wide base to a tall, slender peak. Still other volcanoes are composed wholly of explosion debris. This type of volcano is likely to grow very rapidly, and it usually builds a cone with even slopes.

It may take a million years or more to build a giant volcano of the Hawaiian type, such as Mauna Loa, or one of the composite variety such as Mount Shasta. The viscous kind grows much faster; the steep dome at the top of Mount Pelée, for instance, mushroomed to a height of 1,300 feet within 18 months. But the speed of growth of explosive volcanoes is even more spectacular. The young Mexican volcano of Parícutin was 1,200 feet high on its first anniversary. Monte Nuovo, which grew on the edge of the Bay of Naples in 1538, rose to a height of 440 feet in one day. The record goes to a volcano that sprang up suddenly from Blanche Bay on the island of New Britain in 1937. It attained a height of no less than 600 feet within the first 24 hours; when it stopped growing several days later, it was 742 feet high.

Plateaus

Volcanoes of the kinds we have been considering so far are all made by the discharge of material through a more or less cylindrical conduit in the earth's crust. Such a discharge generally pro- duces a cone, a dome or a sharp, slender spine. But there are also volcanoes in which the magma issues from long fis- sures in the crust. In that case the flood of lava or ash usually produces a plateau, the nature of which depends on the com- position of the escaping magma. There are two general kinds of magma, as every geology student knows. One is the type represented by basalt—a dark, heavy ma- terial, poor in silica and rich in lime, iron and magnesia. Basalt is believed to form a deep-seated, world-wide layer under the oceans and continents. The other is a lighter material, rich in silica and alka- lies; its most typical variety is rhyolite. A basaltic magma is usually hotter and more viscous than a rhyolitic one.

Between 10 and 20 million years ago colossal eruptions of basaltic lava poured out of a region of fissures in the Pacific Northwest. There was a series of erup- tions, sometimes separated by long quiet intervals, so that soils and even forests grew on one flow before being buried by the next. All together some 100,000 cubic miles of fluid lava erupted from the earth and spread over the surface; flow piled on flow until what had been a mountainous terrain was completely bur- ied by a plateau of lava more than 5,000 feet thick and about 200,000 square miles in extent.

The rhyolitic type of fissure eruption, on the other hand, is exemplified by one that took place in 1912 in the Valley of Ten Thousand Smokes in Alaska. In that year swarms of cracks suddenly opened on the valley floor, and a gas-charged, effervescent magma foamed to the sur- face. It was loaded with droplets and clots of incandescent liquid, which cooled to fragments of cellular glass and lumps of white pumice. So mobile was the mixture that it poured for long dis- tances down the valley in the form of glowing avalanches. Since then many other examples of such deposits have been discovered in this country, notably in Nevada and Utah, on the Yellowstone Plateau, in the Globe district of Arizona,

LASSEN PEAK in California has much steeper sides than Mauna Loa. This type of volcano results from eruptions of more viscous material. Lassen Peak was last active in 1917; the cinder cone in foreground, in 1851.

and in the Sierra Nevada and Owens Valley of California. Fissure eruptions of this kind often cause a sinking and downbending of the earth's crust; they account for some of the largest volcanic basins in the world, including those that hold the beautiful lakes of Taupo in New Zealand, of Toba in Sumatra and of Ilopango in El Salvador.

Calderas

One of the most impressive volcanic structures is the type known as a caldera. Calderas are huge pits, shaped like a crater but much larger, usually several miles across. They are also made in a very different way. A crater is the opening through which a volcano discharges its products; it is built during the construction of the cone. A caldera, on the other hand, is a product not of construction but of collapse, for it is created by the cave-in of a crater's sides. In other words, few large volcanoes blow their heads off; usually they are decapitated by engulfment of their tops.

What brings about such a collapse? In composite volcanoes—those built partly of flows and partly of exploded fragments—tremendous explosions of pumice and ash may disembowel the cone and remove support for the volcano's top. The walls of the crater at the summit then founder into the depths. The majestic Crater Lake of Oregon was formed

in this way. A 12,000-foot peak which we now call Mount Mazama once stood there. Some 6,500 years ago volcanic eruptions blew 10 cubic miles of pumice out of its subterranean feeding chamber, leaving a caldera six miles wide and 4,000 feet deep. In the gigantic explosion of Krakatoa in 1883, which expelled some four and a half cubic miles of pumice, the tops of the old volcanoes foundered into the ocean. This produced a caldera five miles wide and propelled a catastrophic tidal wave that drowned 36,000 people on the adjacent coasts of Java and Sumatra.

On the present site of Vesuvius there once stood a much higher volcano. It had lain dormant for so long that vineyards extended to the summit. During this long interval of rest gas pressure accumulated in the underlying magma-chamber. In A.D. 79 it suddenly found release in a succession of terrific explosions. First the lighter, gas-rich head of the magma-column was expelled as showers of white pumice. These buried the town of Pompeii. Then came the debris of a heavier and darker magma from lower levels of the feeding chamber. This clinkerlike material, water-soaked from heavy rains, swept down the mountainsides as mud-flows and demolished the town of Herculaneum. During these violent but short-lived eruptions so much magma was emptied from the volcano's reservoir that the top of the mountain col-

lapsed, leaving a huge, semicircular amphitheatre. Today the wrecked volcano is called Monte Somma; Vesuvius is the younger cone that has risen from the floor of its caldera.

Travelers through the Southwest have seen the giant stone towers that beautify the landscape of the Navajo Reservation, and many have wondered how they were formed. They are actually "embryonic" volcanoes that resulted from short-lived, powerful eruptions. Sometimes gas pressure generated by a body of magma thousands of feet underground blasts cylindrical passages to the surface and hurls out pulverized rock. Such eruptions form explosion pits surrounded by low rims of debris. Erosion may gradually destroy the cones and reveal the cylinders, or pipes, through which the material was fed to the surface. If the explosion material within the pipe offers more resistance to erosion than does the enclosing rock, a time comes when the pipe is left standing alone as a more or less cylindrical tower. Alternatively, if the filling in the pipe is less resistant than the walls, a circular basin persists at the surface. In the Swabian Alps of Southern Germany no fewer than 125 of these embryonic volcanoes occur within an area of about 200 square miles. The same phenomenon is responsible for the famous diamond pipes of South Africa.

Occasionally strong gas explosions

CRATER LAKE in Oregon is a classic volcanic caldera. On this site once stood a 12,000-foot volcano. About 6,500 years ago eruptions expelled 10 cubic miles of pumice from beneath the volcano, which then collapsed into its own feeding chamber (*see opposite page*). Wizard Island, the small body of land that appears in this photograph on the far side of Crater Lake, is a cinder cone that was built up later on the floor of the caldera.

take place at such great depth that the pressure is not strong enough to drill a passage to the surface. Though these muffled "cryptovolcanic" explosions expel no lava or ash, they do deform the earth surface, generally producing a circular depression from two to four miles wide with a central body of uplifted rocks. Walter H. Bucher of Columbia University, our chief authority on cryptovolcanoes, has investigated these peculiar structures in several localities in the U. S., notably at Jeptha Knob, Ky., the Serpent Mound, Ohio, and the Wells Creek Basin of Tennessee.

Steam Explosions

In many volcanic eruptions ground water plays an important part, for the sudden contact of ground water with rising magma produces steam and a violent explosion. This was the cause of a series of strong blasts from the Kilauea volcano in Hawaii in May, 1924. Lava drained from the feed pipes through fissures that opened far down on the sides of the volcano. Many avalanches then tumbled into the pit from the walls and ground water rushed into the empty conduits. The conversion of the water to steam generated enough pressure to blow out the plug of avalanche debris in a series of violent blasts. In 1888 the Japanese volcano of Bandai, which had long been quiescent, erupted with alarming violence. Almost half of the mountain was destroyed and 27 square miles of land was devastated by avalanches resulting from steam blasts that lasted only a few minutes. Presumably ground water had found sudden entry to the hot interior of the dormant volcano.

How much energy is released in a volcanic explosion? Many calculations have been made to determine this, in terms of gas pressures, on the basis of the heights to which eruption clouds ascend and the initial velocities at which bombs are erupted. During explosions of particular violence, such as that of Krakatoa in 1883, fine ash may be blown as high as 30 miles, and even in lesser eruptions it is common to see columns of ash rising 5 to 10 miles into the upper air. In most eruptions the gas pressures range from 50 to 400 atmospheres. In the violent volcano Asama in Japan, which sometimes hurls out large bombs with a muzzle velocity of more than 800 feet per second, the gas pressures may approximate 600 atmospheres.

We have noted that the nature of an eruption depends largely on the viscosity of the magma. The viscosity in turn depends on the magma's composition, its temperature and the amount of gas it holds. The most important factor in producing eruptions probably is the gas. Without gas a magma becomes inert; it can neither flow nor explode. Once the magma, impelled by its relative lightness, has risen from the depths, it reaches a level not far below the surface where the major role in its further advance is played by the effervescence and expansion of bubbles of gas.

The Powerful Charge

What is this gas, this "eruptive element *par excellence*"? In order of importance the gases originally present in the magma seem to be hydrogen, carbon monoxide and nitrogen, with lesser amounts of sulfur, fluorine, chlorine and other vapors. But in the cloud of gas that emerges from a volcano well over 90 per cent is water vapor, with carbon dioxide next in abundance. How much of this water vapor is due to oxidation of hydrogen in the magma, how much is ground water and how much is derived from water-bearing rocks surrounding the magma reservoirs at depth is unknown. Some idea of the prodigious quantities of gas given off from some volcanoes may be gained from the fact that long after the glowing avalanches covered the Valley of Ten Thousand Smokes in 1912, the deposits of pumice continued to give off steam at the rate of six million gallons per second and discharged into the atmosphere some one and a quarter million tons of hydrochloric acid and 200,000 tons of hydrofluoric acid per year.

Apparently gases are important in maintaining high temperatures in magma, in keeping volcanoes alive and in awakening those that are dormant. But this is a speculative matter on which we have little information. The volcano-furnace may be kept hot by the burning of combustible gases; it may also be fueled by other heat-yielding reactions.

At all events, it is the sudden release of gas that accounts for the violent eruptions of long-dormant volcanoes. The gas may be held in solution in viscous magma until heat-yielding reactions near the surface make it boil at an accelerating and finally at a cataclysmic rate; this was the way Mount Pelée discharged the glowing avalanches that destroyed the town of St. Pierre and its 30,000 inhabitants in 1902. Sometimes gases may rise slowly to the top of a magma-column during long intervals of quiet until they either melt or blast an opening to the surface. The spectacular fountains of fiery lava that gush for hundreds of feet into the air during the opening phases of most eruptions of Mauna Loa bear vivid testimony to this upward concentration of gas in a magma-column.

The activity of Vesuvius alternates between periods of relative quiet, when it erupts sluggish flows of lava or rhythmically tosses out glowing bombs, and explosions that burst forth with tremendous strength. Vesuvius produced catastrophic eruptions in 1872, 1906, and 1944. During the intervals between these explosions minor eruptions gradually increased the height of the central conelet and therewith the height of the central column of magma. Then the sudden opening of fissures far down the sides of the mountain allowed lava to escape quickly from the lower part of the column. The draining of the column greatly reduced the pressure on the underlying magma, allowed a large amount of dissolved gas to escape suddenly from solution and thereby produced colossal explosions.

Predicting Eruptions

One object of any science, perhaps the chief, is to improve our powers of

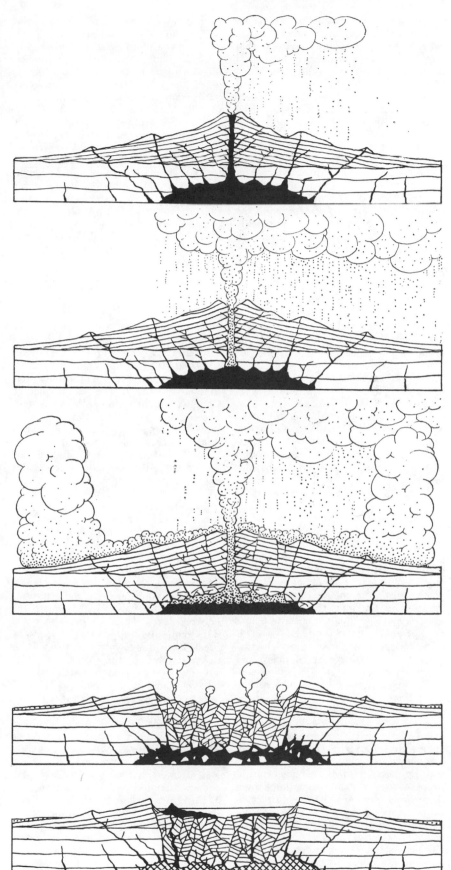

EVOLUTION OF CRATER LAKE presumably proceeded along the lines shown in these drawings. At first magma stood high in the volcano; later it sank as eruptions of pumice increased. Then the volcano collapsed into its chamber. Finally the caldera filled with water and Wizard Island arose.

prediction. Is it possible to say where and when volcanoes may erupt? To some extent, yes. As to where, we know that they are likely to erupt only in regions of recent mountain-building and fracturing, in or close to belts of marked earthquake activity and almost surely only in regions still or lately marked by volcanism. When? That is a more difficult question.

We can get some warning from the seismograph. An increase in the number and intensity of local quakes in a volcanic region is fairly sure to herald an eruption. For 16 years prior to the great eruption of Vesuvius in A.D. 79 the neighboring region was repeatedly shaken. For 20 days before Parícutin was born in Mexico in 1943 the surrounding country trembled from increasingly numerous and vigorous shocks. T. A. Jaggar and R. H. Finch of the Hawaiian Volcano Observatory have foretold when Kilauea and Mauna Loa would erupt, by study of the distribution of quakes caused by fissuring of the ground as magma surged toward the surface. In many volcanic regions such preliminary quakes are accompanied by subterranean rumblings and by avalanches from the walls of craters.

Next to seismic evidence, tilting of the ground around dormant volcanoes is perhaps the most reliable clue to impending activity. The underground movement of magma often causes rapidly changing tilts on the surface. Indeed, active volcanoes almost seem to breathe; they are forever swelling and subsiding as the subterranean magma fluctuates in level. By combining strategically placed tiltmeters and seismographs, it has been found possible to say approximately where, as well as when, an eruption of Mauna Loa will take place. Accurate measurement of the cracks along the rim of the Kilauea crater also serves as a guide, for these cracks are not just superficial openings caused by slippage but mark fundamental planes of weakness that go deep, and when they widen rapidly it usually means that magma is rising beneath the crater floor.

Another hint of imminent eruption may be given by strong local disturbances of the earth's magnetism. These are produced by the rise of hot, nonmagnetic magma in the volcanic pipes and by heating of the adjacent wall-rocks. Along with the magnetic changes there are commonly changes in electrical currents in the earth; these are detected, for instance, several hours before every violent explosion of the Japanese volcano Asama. Other indications of impending eruptions, though rather unreliable, are a sudden rise in the temperature of hot springs and gas vents around a dormant volcano, a pronounced increase in the volume and acidity of escaping gases, or a sharp increase in the temperature of near-surface rocks.

Some volcanoes behave in a roughly cyclic fashion, so that the likely sequence of events may be foretold. For instance, when the central conduit of Vesuvius has grown to unusual height, the danger of a flank outburst of lava followed by catastrophic explosions from the summit is at a maximum. And once an eruption has begun, it may be possible to predict fairly well what is likely to follow. Thus the late Frank Perret of the Carnegie Institution of Washington, by a careful analysis of the early phases of an eruption of Mount Pelée in 1928, was able to reassure the frightened inhabitants of St. Pierre that there would be no repetition of the awful calamity of 1902.

Volcano Control

Given sufficient warning, it is sometimes possible to minimize the damage caused by a volcano's eruption. The first recorded effort of this kind was undertaken during an eruption of Mount Etna in Sicily in 1669. The inhabitants of Catania, a town in the path of the lava pouring down the mountain, made a brave attempt to save their city by digging a channel to divert some of the lava. Unfortunately the new stream moved toward a neighboring town, the angry citizens of which soon put a stop to the efforts of the Catanians.

In recent years the U. S. Air Force has tried the experiment of checking lava flows by bombing them from the air. These tests were made on the Mauna Loa volcano in Hawaii. There are three ways in which a flow may be checked. The method of attack depends on the stage or circumstances of the eruption. Where the lava flows as an open stream between levees of partly crusted material, bombs are dropped on the levees to break them down. Some of the lava is diverted through the breaks, the pressure on the main stream is relieved and the flow may then come to a halt. Where the lava has become completely crusted over, so that the liquid portion pours through internal tubes, bombing may not only produce breaks permitting it to escape but may clog the tubes with solid debris and cause enough stirring and effervescence of gas so the liquid becomes more viscous or even congeals. Thirdly, bombardment may break down the walls of the cinder cone at the source of the flow itself, thus dissipating the energy of the eruption in many minor flows.

In Java dams have been built to divert volcanic mudflows away from villages and agricultural lands, and for some of the more active volcanoes there are danger maps indicating areas particularly liable to devastation. Installed on the walls of a canyon on the side of the volcano Merapi are thermoelectric devices which cause an alarm to ring in the volcano observatory when the air in the canyon is heated by passing lava or hot avalanches; villages in the danger zones can then be warned. In addition there are artificial hillocks in some of the villages which serve as islands of refuge from volcanic mudflows.

Harnessing the Energy

Naturally a good deal of thought has been given to how the immense energy of volcanoes might be harnessed for man's use. It has been done on a relatively minor scale in several countries, notably Italy and Iceland. In Iceland many buildings are heated by volcanic steam, and by warming fields with steam pipes the country is able to raise crops that normally grow only in more temperate climates. In Italy natural steam has been used to generate electricity since 1904. There is a region in Tuscany where steam from a deeply buried body of magma comes out of the ground through rifts and is also tapped artificially by means of wells. A typical well develops a pressure of about 63 pounds per square inch, and it yields 485,000 pounds of steam per hour at a temperature of about 400 degrees Fahrenheit. In 1941 Tuscany's harnessed volcanic steam generated 100,000 kilowatts of electric power. In addition, a large amount of boric acid, borax, ammonium carbonate, carbon dioxide and other chemicals was recovered from the vapors.

The energy available from the gas vents and hot spring waters of volcanic regions is of fantastic proportions. The hot springs and geysers of Yellowstone National Park, for instance, are calculated to give off 220,000 kilogram-calories of heat—enough to melt three tons of ice—every second. A well drilled to 264 feet in the Norris Basin developed a steam pressure of more than 300 pounds per square inch. At "The Geysers," 35 miles north of San Francisco, there are wells which, it is estimated, could provide an average of more than 1,300 horsepower each.

Thus far little use has been made of this available energy. There are many technical difficulties, of course, in the way of large-scale utilization of volcanic power, not least among them being the acidity of many of the vapors. But one can expect with confidence that these difficulties will be largely overcome, and that more widespread use will be made of the stores of energy now running to waste.

ATITLÁN AND TOLIMAN VOLCANOES in Guatemala are examples of composite cones formed both by the outpouring of lava and the explosive discharge of rock. On the sides of the cones are flows from fissures.

MONO CRATERS in California are plugs of solidified lava surrounded by pumice. In the southwestern U. S. the erosion about volcanic plugs has proceeded to such an extent that they are left standing as stone towers.

9

The Flow of Heat from the Earth's Interior

by Henry N. Pollack and David S. Chapman
August 1977

A global heat-flow map can be drawn on the basis of thousands of individual field measurements on continents and ocean floors. The heat-flow pattern is interpreted in terms of plate tectonics

"Straight through the dim and open portal we entered unopposed, and I, eager to learn what part of Hell's bowels those burning walls enclosed, began to look about." So did Dante write of his descent into the Inferno. Miners of later centuries might well have believed he was describing their daily working environment, since it has been widely observed that without proper ventilation and cooling a mine gets hotter with depth. There are of course many other indications that the earth's interior is hot, the most obvious being an erupting volcano. Only slightly less spectacular are the handful of areas around the world that display hydrothermal activity, such as the hot springs, steam vents and geysers of Yellowstone National Park. One of the fundamental axioms of physics, embodied in what is known as Fourier's law of heat conduction, is that heat flows from the warmer parts of a body to the cooler ones. It can therefore be inferred that since the temperature increases with depth in the earth's crust, there is a flow of heat outward from the earth's interior.

The transfer of heat within the earth and its eventual passage to the surface by conduction through the crust play a fundamental role in all modern theories of geodynamics. In the 19th century the earth's internal heat also figured significantly in the protracted debate over the age of the earth between William Thomson (Lord Kelvin) and several of his scientific contemporaries. Kelvin's dissertation at the University of Glasgow in 1846, titled "Age of the Earth and Its Limitations as Determined from the Distribution and Movement of Heat within It," was the first of a long series of papers in which he laid out the argument that the earth's thermal gradient (the rate at which the temperature increases with depth) would continue to diminish with time as the earth cools following its formation and solidification from mol-

ten rock. By determining the earth's thermal gradient from measurements in mines and boreholes, he maintained, one could tell how long the earth had been cooling and so could determine the age of the earth. Records of temperatures at various depths in mines could be found in mining journals, and Kelvin supplemented them with measurements of his own in Scotland. His conclusion was that the temperature, at least to modest depths below the surface, increased at a rate of between 20 and 40 degrees Celsius for every 1,000 meters of depth. To Kelvin this relation indicated that the earth had been cooling for only a few tens of millions of years, a period far shorter than many geologists and biologists of the time thought necessary for the development of the known stratigraphic and fossil record. The ensuing debate spanned half a century and pitted Kelvin against such prominent evolutionists as Charles Darwin and Thomas Huxley.

Kelvin's calculation was based on the assumption that the heat being lost by the earth was drawn from the reservoir of heat left over from the earth's originally molten condition. That assumption, which was essentially unchallenged for decades, was to be the undoing of all estimates of the earth's age based on the measurements of its heat. Three observations made Kelvin's estimate of the age of the earth based on its initial heat no longer tenable: the discovery of radioactivity by Henri Becquerel in 1896, the observation by

Pierre Curie in 1903 that the radioactive decay of certain isotopes liberates heat and the confirmation by Robert Strutt in 1906 that common rocks found in the earth's crust contain sufficient amounts of radioactive isotopes to yield a significant fraction, if not all, of the earth's observed heat flow.

Measuring Heat Flow

How much heat is the earth losing today as a result of conduction from its interior? The global average is close to .06 watt per square meter of surface, or about 30 trillion watts over the entire planet. The amount of energy arriving from the sun is almost 6,000 times greater, and it is completely dominant in establishing the temperature of the earth's surface. The flow of heat from the interior is scarcely a trickle; the heat conducted through an area the size of a football field is roughly equivalent to the energy given off by three 100-watt light bulbs. The evolution of the earth covers vast reaches of time, however, and a trickle of energy over aeons can do significant geological work, such as making continents drift, opening and closing ocean basins, building mountains and causing earthquakes. The geographic variation in the flow of heat from the earth's interior is not great: most measurements lie within a factor of three around the mean value. The patterns of heat flow in continental regions differ from those in oceanic ones, but the average heat flow through both

DRAMATIC EVIDENCE of the power of the earth's internal heat to mold the geology of the surface is provided by this photograph of the 1971 Mauna Ulu eruption on Kilauea volcano in Hawaii. The photograph was made by Wendell A. Duffield of the U.S. Geological Survey as he was standing on the rim of the Mauna Ulu crater and looking almost straight down. The bluish gray background is the comparatively cool crust of partly solidified basaltic lava that forms on the surface of the hot liquid-lava "lake" below. The jagged orange streaks are cracks through which molten lava is upwelling. The entire scene, Duffield points out, is analogous to the spreading of new sea floor from mid-ocean ridges visualized in the plate-tectonic model.

DRILL RIG IN ZAMBIA, originally set up to bore into the earth in search of copper, provided one of several "holes of opportunity" used by the authors and their co-workers as part of their program to obtain heat-flow measurements in areas of Africa and South America where the existing data are sparse. The earth's temperature is taken by lowering an electrical-resistance thermometer (called a thermistor) down the borehole, making measurements at several depths. The records are used to establish the rate at which the temperature of the rock increases with depth, a local quantity known as the geothermal gradient. The Zambian heat-flow measurements were carried out four years ago while the authors were on leave from the University of Michigan and were based at the University of Zambia. More than 50 such drill holes were surveyed at eight different Precambrian geological sites in the country. The results of the survey were interpreted by the authors as indicating the presence of anomalously warm material only a few tens of kilometers below the surface of the earth.

is surprisingly similar. Some areas, such as Iceland, exhibit an extraordinary heat flow, and geothermal areas of this type can be tapped as an energy resource.

If heat is being transported through the earth's crust by thermal conduction, the amount of heat in transit is equal to the product of the temperature gradient times the thermal conductivity (a property of the rock that describes how easily it transmits heat). Any experimental study of the earth's heat flow is concerned with measuring these two quantities. On continents temperature gradients are measured by lowering sensitive electronic thermometers—thermistors—down drill holes or by measuring the temperature of the rock at different levels in mines. The process of drilling a hole disturbs the thermal equilibrium at the site; hence several weeks or months are allowed to lapse between the drilling and the measuring. Even after the disturbance has become negligible, subsurface temperatures are disturbed by such effects as the daily and annual fluctuation in the surface temperature, unevenness in vegetative cover, unevenness in topography, the movement of ground water, the uplift or erosion of the surface and variations in climate. Most of these disturbances diminish to an acceptable level beyond depths of a few tens of meters; some, however, can extend to several hundreds of meters. Although reliable heat-flow measurements can sometimes be made in holes as shallow as 50 meters, most workers who make such measurements prefer to do so in holes that are 300 meters or more in depth.

On the ocean floor, where sediments are comparatively soft and the blanket of seawater provides an environment of almost constant temperature, the drilling of a hole is unnecessary. There temperature gradients are determined by plunging a long cylindrical probe several meters into the soft sediment and measuring the temperature at one-meter intervals with fixed thermistors.

For measurements of thermal conductivity two methods are widely used. For hard continental rock a sample from the drill hole is cut and polished in the form of a disk and is inserted into a column between silica disks of known conductivity. A constant temperature difference maintained between the ends of the column causes a flow of heat through the sample and the silica standards, and the measurement of the relative drop in temperature across the components of the stack yields the thermal conductivity of the sample. For softer continental rocks and marine sediments a thin needle is inserted into the sample and is heated along its length. From a record of the rise of temperature with time the thermal conductivity can be calculated easily.

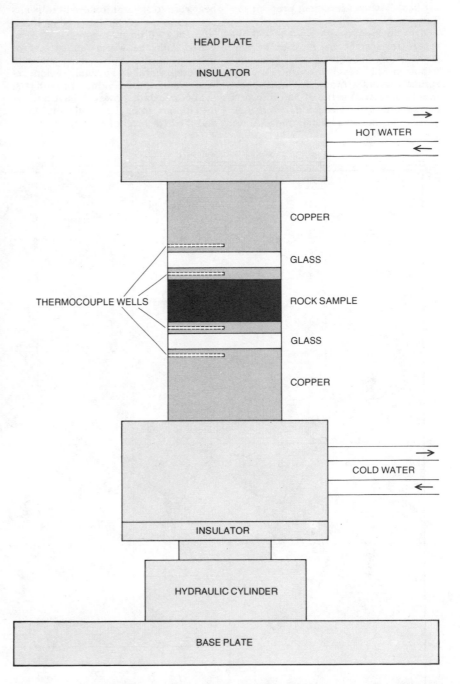

THERMAL CONDUCTIVITY of a hard rock sample obtained from a drill hole is measured in the apparatus depicted here. The sample, cut and polished in the form of a disk, is inserted in a column between silica disks of known conductivity. Constant temperature differences maintained between the ends of the column cause a flow of heat through the sample and the silica standards. The sample's thermal conductivity is determined by measuring the relative drop in temperature across the components of the stack. The heat flow at the drill site is equal to the product of observed local temperature gradient times thermal conductivity of the rock.

More than 5,000 such heat-flow measurements have been reported in a recent compilation by Alan M. Jessop of the Canadian Department of Energy, Mines and Resources and by John G. Sclater and Michael A. Hobart of the Massachusetts Institute of Technology. Although the number of measurements is sufficient for several types of regional analysis, the data set is still geographically uneven: more than twice as many measurements have been made at sea as on land. The middle-latitude oceans, North America, Europe and Australia are quite well surveyed, whereas large areas of the high-latitude oceans and of South America, Africa, Asia and Antarctica have no measurements at all. In the past four years our group from the University of Michigan has helped to remedy some of this geographic imbalance on the continents by conducting heat-flow measurement programs in Zambia, Niger and Brazil.

An analysis of the global data without regard for specific site location or geologic setting indicates a fairly wide distribution of results asymmetrically spread around a modal (or most commonly observed) value of 50 milliwatts per square meter [*see top illustration on page 118*]. Individual values range from near zero to several hundred milliwatts per square meter, the latter being located mainly within geothermal areas associated with the worldwide system of mid-ocean rifts. Subdividing the data into continental and oceanic regions reveals similar modal values for both sets; the oceanic data, however, have a wider distribution than the continental data, and the high level of asymmetry forces the means, or average values, for each of these regions well above the modes.

This gross grouping of heat-flow measurements has been useful in the past, and the similarity between continental and oceanic measurements has served to stimulate much discussion among geophysicists and geologists who had expected quite different results in the two settings. As with many other aspects of

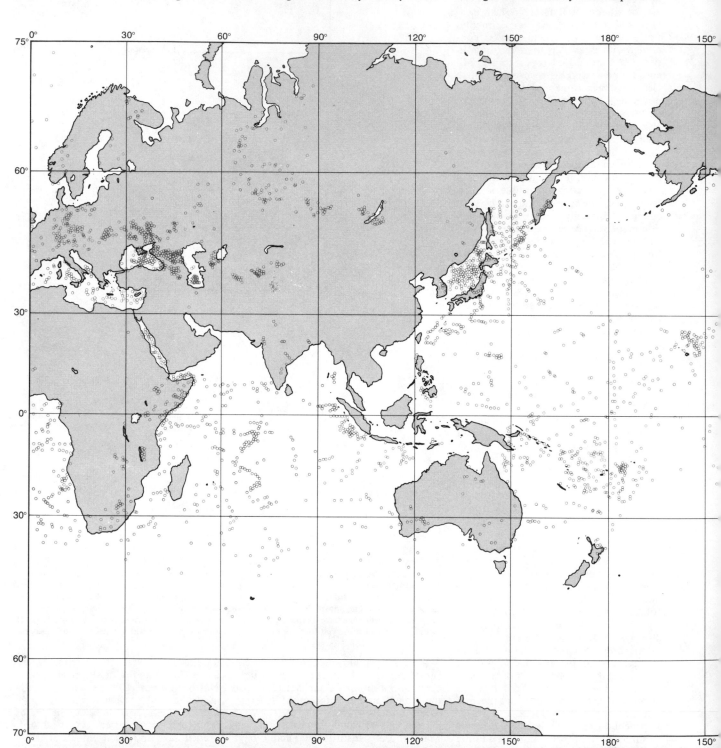

UNEVEN GEOGRAPHIC DISTRIBUTION of oceanic and continental heat-flow measurements (*colored circles*) is evident on this world map, which is based on one prepared by the National Geophysical and Solar-Terrestrial Data Center. The number of such measure-

earth science, however, heat-flow observations now find a compelling new interpretation in terms of the concepts of sea-floor spreading and plate tectonics.

Heat Flow and Plate Tectonics

According to plate tectonics, the lithosphere, the outer shell of the earth, is made up of a dozen or so rigid plates that are being moved about on the earth's surface. Wherever plates are moving apart the gap is filled by hot material flowing upward from the earth's underlying mantle. This material accretes to the edges of the separating plates; the accreting edges form the mid-ocean ridges. The new rock cools as it moves away from the ridge. Across the plate from a ridge one usually finds a great oceanic trench, which marks a site where older and cooler plate is subducted, or recycled back into the interior. Frictional and conductive heating of the plate in the subduction zone melts part of it, and the melted fraction rises buoyantly to the surface to form the volcanoes and island arcs typically arrayed behind the trenches. Such subduction processes, together with other forms of plate interactions, give rise to thermal metamorphism, the generation of volcanic magma and mountain building on continents.

One first looks to the oceanic plates with their comparatively simple geology to obtain evidence for the thermal model of plate evolution. Edward C. Bullard of the University of Cambridge, who reported the first marine heat-flow measurements for the Atlantic in 1954, noted at that time the near-equality of the mean heat flow from the continents and the ocean floor. Today, with the addition of some 3,500 measurements in ocean floor of all ages, it is possible to see a systematic decrease of heat flow with increasing age and hence depth [*see bottom illustration on next page*]. For those sites where a thick, impermeable cover of sediments prevents the removal of heat by seawater circulating through the fractured oceanic crust, heat-flow measurements agree extremely well with predictions based on mathematical models of a cooling plate.

Such models of plate cooling also explain the broad topographic features of the ocean floor. The newly formed crests of mid-ocean ridges are typically 1,000 to 3,000 meters below sea level, whereas the oldest ocean basins are 5,500 meters below sea level. Thus in 200 million years the sea floor subsides by about 3,000 meters. The reason is that the recently accreted rock is hot and thermally expanded, whereas the older material has cooled and contracted. The match between the observed topography and the theoretically calculated topography is so good that it seems unlikely the subsidence will be explained in any way other than in terms of a simple cooling model.

The concept of the thermal evolution of an oceanic plate may also provide an answer to a long-standing puzzle in geology: What gives rise to transgressions of the sea onto continents? The Upper Cretaceous period was one such time of great marine transgression. Starting about 100 million years ago the sea level rose; it crested between 90 and 70 million years ago and withdrew from the continents about 60 million years ago. From the spacing of magnetic lineations on the ocean floor it can be shown that the Upper Cretaceous was also a period of rapid sea-floor spreading. Since the cooling and subsidence of an oceanic plate is time-dependent, an increase in

ments has increased rapidly in recent years: from 47 in 1954 to 1,162 in 1962 to more than 5,000 at present. More than twice as many measurements have been made at sea as on land.

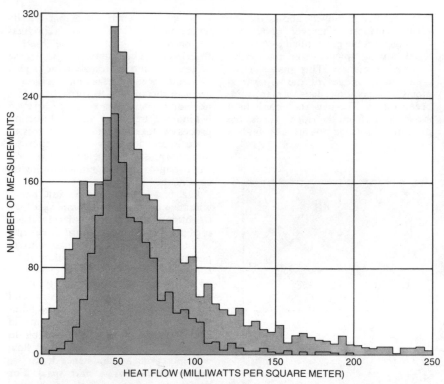

STRIKING SIMILARITY is seen in the asymmetrical distribution of heat-flow values for both continents (*color*) and oceans (*gray*). Most of the observed values are in the range between 20 and 120 milliwatts per square meter, with the global average being at about 60. The modal, or most commonly observed, values for continents and oceans (*peaks of profiles*) are both closer to 50. It is not known at present whether the near-equality of the continental and the oceanic heat-flow measurements is a fundamental characteristic of the movement of heat within the earth or is merely a coincidence arising from incomplete sampling. The data for this chart were compiled by Alan M. Jessop of the Canadian Department of Energy, Mines and Resources and by John G. Sclater and Michael A. Hobart of the Massachusetts Institute of Technology.

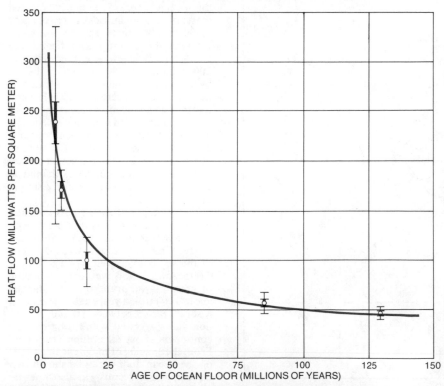

HEAT FLOW DECREASES with increasing age of ocean floor, as indicated here by the distribution of measured heat-flow values for five selected areas on the floor of the Pacific Ocean. Data points give the mean value for each age grouping. Heavy bars show the probable error; light bars show standard deviation about the mean. Measurements agree extremely well with a theoretical estimate of heat flow expected from a cooling plate of oceanic crust (*colored curve*).

the spreading rate would have broadened the oceanic ridge and increased its volume. This in turn would have reduced the water capacity of the ocean basins and displaced the sea onto the continents. The subsequent regression was apparently caused by a reduction in the rate of sea-floor spreading that began about 85 million years ago.

Above subduction zones the heat-flow patterns are more complex, but they nonetheless provide important clues to the subduction process. A pattern generally observed at subduction zones, and particularly well documented for the Japan arc system, is one of low heat flow near the oceanic trench and very high heat flow to the landward side of the island arc [*see illustration on opposite page*]. The pattern suggests that the top part of the cool subducting plate acts as a heat absorber, causing the band of low heat flow observed adjacent to the trench. Deeper in the subduction zone frictional and conductive heating are sufficient to melt part of the plate, yielding as a product the volcanic island arc itself and the augmented heat flow behind the arc.

The western U.S. provides another example of an elongated zone of low heat flow adjacent to a region of magmatic activity and high heat flow [*see illustration on page 120*]. Here, however, there is at present no active major subduction zone nearby. David D. Blackwell of Southern Methodist University and others have suggested that this zone represents a fossil heat-flow pattern, established some tens of millions of years ago when subduction was active along the boundary between the Pacific plate and the North American plate.

Continental heat flow in areas removed from plate boundaries also falls into recognizable patterns. Measurements on continents now number about 1,700, and from these data one can draw two major conclusions. First, there is a general decrease in heat flow with the increasing age of a geologic province [*see top illustration on page 121*]. This result is similar to that for oceans, but the time scale is apparently quite different. Whereas oceanic heat flow drops below 50 milliwatts per square meter after 100 million years of cooling, on continents one finds such heat flow in geologic settings four or five times older.

The second major result is that for large areas of continents there is a clear relation between surface heat flow and the radioactivity of the surface rocks. That continental rocks, granites in particular, generate significant quantities of heat by the spontaneous disintegration of radioactive elements has been known since early in this century. In 1968 A. Francis Birch, Robert F. Roy and Blackwell, all then working at Harvard University, demonstrated that when heat-flow measurements are plotted with respect to radioactive heat generation for the rocks at various sites, the plotted values fall along a straight line [*see bottom illustration on page 121*]. Different lines were obtained for the eastern U.S., the "basin and range" geologic province of Nevada and Utah, and the Sierra Nevada region, but within each region a linear relation holds. This finding implies that for a given region the heat flow at the surface has two components: a crustal component that varies from site to site according to the local radioactivity and a deeper component that originates in the earth's mantle and is uniform for all sites in the region. Since 1968 this relation has received much attention. Arthur H. Lachenbruch of the U.S. Geological Survey has looked into its consequences for the distribution of heat-producing isotopes in the earth's crust, and he has explained why the concentration of such isotopes should be expected to diminish exponentially with depth.

The variation of the mantle-derived component of the observed heat flow between different provinces has been less well studied, but on the basis of the limited data available we have speculated that in most heat-flow provinces there is a regular partitioning of the heat flow, with about 40 percent of the mean surface flux coming from within the zone of crustal enrichment and 60 percent coming from below. This partitioning, if confirmed, suggests that the average heat production of the continental crust should vary inversely with its age, because in general the older provinces display less heat flow. Such a relation can be explained with a model in which radioactivity diminishes with depth, on the assumption that the older geologic provinces have been eroded to greater depths than the younger ones.

Another intriguing finding has been reported recently by Tom Crough of Stanford University and V. M. Hamza of the University of São Paulo. They show that when one subtracts the heat-flow contribution originating in the zone of crustal enrichment, the remaining heat flow continues to show an inverse dependence on the age of the province, but the time scale of this residual heat flow appears to be one of a simple cooling process, much like what is observed in the oceans. The cooling on the continents is apparently much further along, however, and it must have penetrated more deeply there. Could it be that we are seeing residual heat from a tectonic event 600 million years ago? If we are, the implication is that such events must involve at least the outer 500 kilometers of the earth in order for any residual heat to be making its way to the surface today.

Global Heat Flow

Let us now turn our attention to the broad features of the thermal field of the

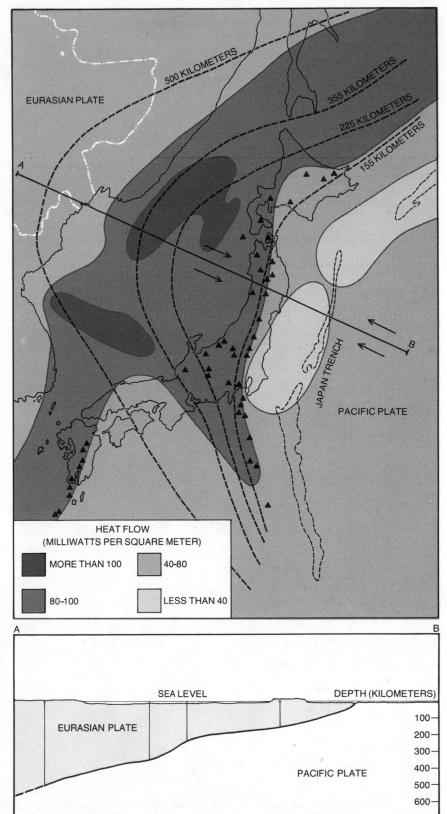

DISTINCTIVE HEAT-FLOW PATTERN is produced by the subduction of the tectonic plate underlying the Pacific Ocean as it dives under the islands of Japan. The arrows in the map at the top indicate the relative convergence of the Pacific and the Eurasian plates, and the broken lines show the depth to the subducted slab. The corresponding depths are also indicated in the cross-sectional diagram at the bottom, which is drawn approximately to scale along the line *AB*. The low heat-flow zone (*lightest color*) observed between the Japan trench and the island arc suggests that from the surface to a depth of about 120 kilometers the cool subducting plate acts as an absorber of heat from the earth's mantle. The volcanoes of the island arc (*black triangles*) and the region of high heat flow (*darkest color*) between Japan and mainland Asia result from frictional heating and partial melting deeper in the subduction zone.

entire earth and in doing so combine results from both the continental and the oceanic regions. Before 1974 several attempts had been made to plot the observed variations in heat flow on a global scale. In spite of the growing number of heat-flow measurements, however, there were still large areas of the globe where no data had been gathered. A mathematical representation of the global distribution of heat-flow measurements was desirable in order to correlate regional heat flow with other geophysical phenomena, such as the earth's gravitational field. The analysis was beset with difficulties because of the need for extensive extrapolation into unsurveyed areas.

By 1974, however, the relation between heat flow and age for both continental and oceanic regions was well established. Could these known correlations not be utilized to make estimates of the probable heat flow in unsurveyed areas and so guide the heat-flow contouring on a global map? They could if geological maps were available showing the ages of all continental and oceanic regions. Although such maps had existed for the continents for some time, it was not until 1974 that Walter C. Pitman, Roger L. Larson and Ellen M. Herron of the Lamont-Doherty Geological Observatory summarized on a single map the detailed ages of all the oceanic regions based on magnetic anomalies on the ocean floor and the recorded reversals of the polarity of the earth's magnetic field. Soon after obtaining the map we divided the entire earth into grid elements five degrees on a side and proceeded to assign to each element a heat-flow value based on the relation of heat flow to tectonic age and the fraction of different age groups present in the grid element. In effect we were creating a synthetic estimate of heat flow in unsurveyed areas. The full data set, comprising observations supplemented by estimates, could then be fitted by appropriate mathematical functions and plotted with a minimum of distortion [see top illustration on pages 122 and 123].

The new world heat-flow map constructed in this way showed for the first time on a global scale variations in heat flow that had been recognized in regional surveys. All the major oceanic-ridge systems can be seen as heat-flow highs, as are Alpine Europe, much of western North America and the marginal basins of the western Pacific. The Galápagos spreading center and the Chile Rise appear as bulges on the dominant East Pacific Rise. Regions of low heat flow include all the major continental shields and sedimentary platforms and the oldest oceanic regions.

The Thickness of the Lithosphere

The determination of the thickness of the lithosphere has until recently been a seismological endeavor. The seismologist's method is to observe the dispersive effects of a given region of the earth on earthquake surface waves that propagate across it. The analysis of dispersion patterns yields information about the elasticity of the crust and the upper mantle; in particular the pattern of dispersion can confirm the existence and locate the position of a zone where seismic waves travel at low velocity. Such a zone is probably a manifestation of a region of partial melting in the mantle, and many geophysicists identify this zone as the asthenosphere postulated in the plate-tectonic model. Since the lithosphere is what lies above the asthenosphere, the depth to this low-velocity zone is equal to the thickness of the lithosphere.

It has been known for some time that under the old Precambrian shields of the continents the seismic low-velocity zone either is absent or is deep and only weakly developed. In contrast, the young and active geologic provinces such as those in the western U.S. have a shallow and well-developed low-velocity zone. Recently a number of reports have appeared suggesting that the depth to the low-velocity zone under the oceans also increases with the age of the ocean floor, implying that the oceanic lithosphere progressively thickens with time.

The depth at which partial melting takes place in the mantle in a given region depends on the temperature at which the rock of the mantle begins to melt and on the variation of temperature with depth. The depth profile of the actual temperature, called the geotherm, in turn depends strongly on the heat flow. Thus with the aid of considerable extrapolation surface heat-flow

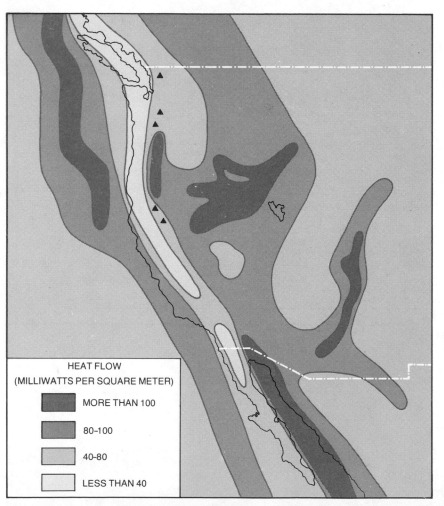

HEAT FLOW (MILLIWATTS PER SQUARE METER)

- MORE THAN 100
- 80–100
- 40–80
- LESS THAN 40

FOSSIL SUBDUCTION ZONE appears to account for the elongated region of low heat flow observed adjacent to a magmatic region of high heat flow in western North America. Sea-floor spreading under way at present in the Gulf of California and at the Gorda and the Juan de Fuca oceanic ridges results in high heat flow offshore. Inland the parallel belts of low and high heat flow mark the shallow and deep parts respectively of a subduction zone that was active during the early Cenozoic era. Although the subduction of the oceanic plates off central California ceased more than five million years ago, a heat-flow pattern similar to that seen in the currently active Japan arc system persists. Black triangles again denote recent volcanoes.

data can be used to predict the thickness of the tectonic plates.

Since direct measurement of temperatures in the earth is limited to the top 10 kilometers of the crust, the extrapolation of temperatures to depths of 100 kilometers or so involves several assumptions. One needs to know how the thermal properties of the rock vary with temperature, how radioactivity is related to depth and for oceanic regions how the oceanic plate cools after it is formed at the ridge. Recent laboratory measurements and field observations have provided enough data for the construction of detailed models, so that one can calculate characteristic geotherms for both continental and oceanic regions with some confidence [*see illustration on page 124*]. The depths to partial-melting conditions predicted from such calculations agree well with the seismologists' results from their surface-wave studies. Both the heat-flow measurements and the seismological data indicate that oceanic plates thicken as they age, from a few kilometers soon after their formation at a ridge to 100 kilometers or more in the oldest ocean basins, where the heat flow is low.

The continental portions of the tectonic plates also show a systematic variation in thickness, from 40 kilometers in young geologic provinces whose heat flow is high to several hundred kilometers under continental shields whose characteristic heat flow is much lower [*see bottom illustration on next two pages*]. For some shields the geotherm does not intersect the mantle's melting curve at any depth, and so in a strict sense the asthenosphere should not exist under the shields. In these areas thick lithosphere would be coupled directly to the deeper interior, acting as an "anchor" to retard the motion of the plate system. More realistically, we expect that the plate constituting a shield does decouple from the deeper interior, probably at a depth at which the geotherm makes its closest approach to the melting curve of the mantle. In a sense, then, the continental shields are at present probably dragging anchor.

Thermal History of the Earth

What can be said about the thermal state of the earth in times long past? Most treatises on this subject begin by telling why the problem is a difficult one and then offer a series of disclaimers in case one or another assumption should turn out to be invalid. (The lesson of Kelvin has not been lost on others!) Nevertheless, some observations can be ventured that make possible certain broad inferences, even though the details necessarily remain indistinct. One first must note the abundance within the earth's crust of the principal heat-pro-

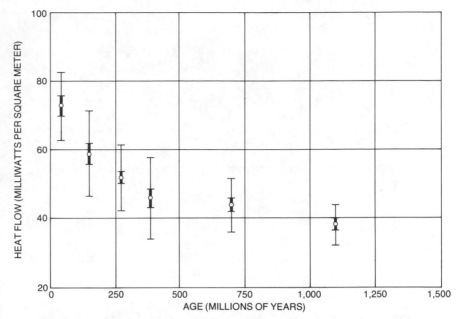

ON CONTINENTS the distribution of heat-flow values also shows an apparent relation to the age of the last major tectonic event that affected the region, just as the distribution of oceanic heat-flow values is associated with the time elapsed since volcanic rock began to cool following the extrusion of mantle material at an oceanic ridge. In the case of continents, however, the measurements indicate that the loss of heat after the tectonic event extends over an interval that is four or five times longer than the decay time for heat flow from the oceanic plate.

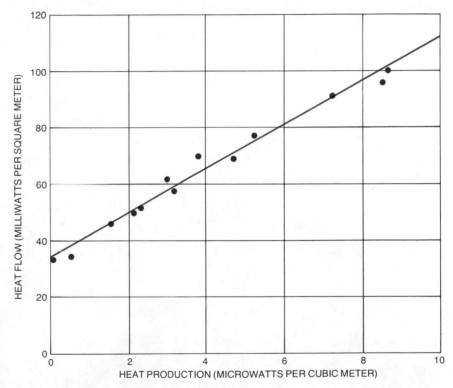

CHARACTERISTIC LINEAR RELATION is frequently discovered in plotting heat flow against heat production for a particular region, as can be seen in this graph for data obtained in the eastern U.S. (*black dots*). Such variations in heat-flow data for different continental regions can often be attributed to differing concentrations of heat-producing radioactive isotopes in the outer few kilometers of the earth's crust. The point where the colored line intercepts the vertical heat-flow axis indicates the amount of heat flow coming from below this zone of crustal radioactive-isotope enrichment. The slope of the line tells how deep within the crust the enrichment persists. The slope and intercept values together serve to characterize different heat-flow provinces. The variation of heat flow within a province is governed by the regional variability of the crustal isotopic enrichment. The differences in heat flow between provinces arise principally from variations in the amount of heat flow coming from below the crust.

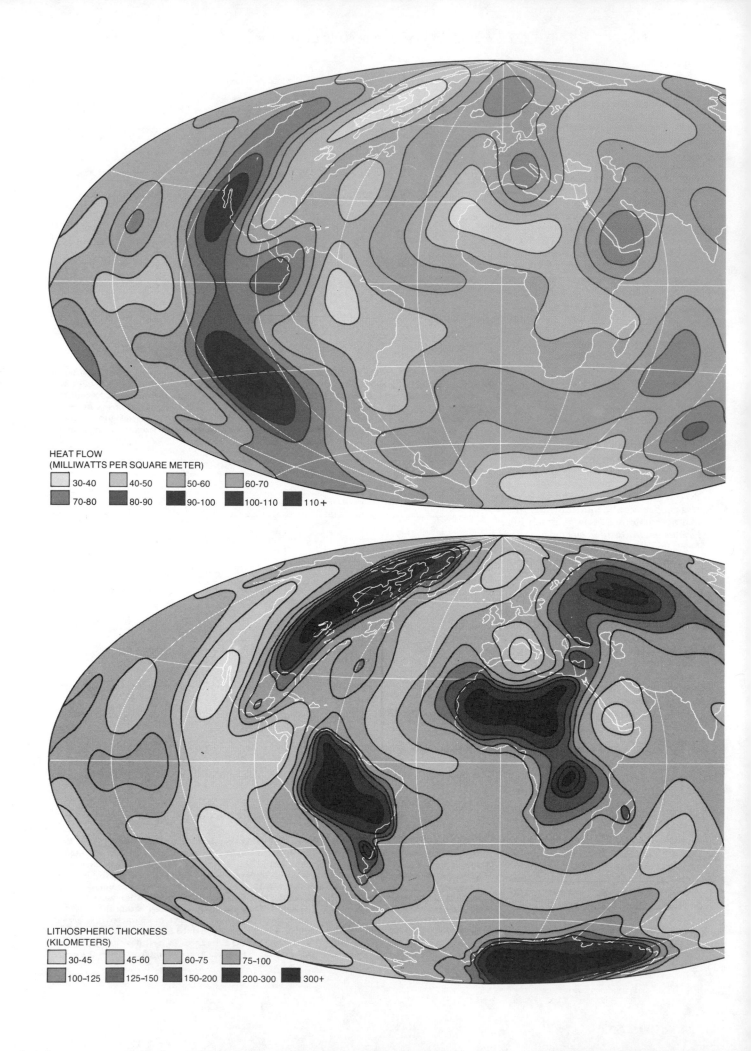

HEAT FLOW
(MILLIWATTS PER SQUARE METER)

30-40 40-50 50-60 60-70
70-80 80-90 90-100 100-110 110+

LITHOSPHERIC THICKNESS
(KILOMETERS)

30-45 45-60 60-75 75-100
100-125 125-150 150-200 200-300 300+

WORLD HEAT-FLOW MAP, constructed by the authors on the basis of the available observations supplemented by estimates, shows variable patterns of heat flow on a global scale. The main zone of high heat flow (*darker shades of color*) is in the eastern Pacific Ocean off Central America and South America. This zone coincides with the East Pacific Rise, a major oceanic ridge where new sea floor is being extruded and carried away quite rapidly to yield a comparatively broad band of high heat loss. The oceanic ridges in the Atlantic Ocean and the Indian Ocean are spreading more slowly and hence result in a narrower zone of above-average heat flow. Other regions of fairly high heat flow include the marginal ocean basins of the western Pacific, which overlie active subduction zones, the western cordillera of North America, where subduction ceased between five and 10 million years ago, and Alpine Europe. The principal regions with below-average heat flow (*lighter shades of color*) include all the ancient Precambrian shield and platform areas of the continents and the oldest parts of the ocean floor.

ducing radioactive isotopes: thorium 232, uranium 238, potassium 40 and uranium 235. The continental crust averages 40 kilometers in thickness, less than 1 percent of the earth's radius, yet its endowment of these heat-producing isotopes is great enough for 40 percent of the heat flow at the earth's surface to arise within the crust itself. The concentration of isotopes in the oceanic crust is less, but it still represents a significant enrichment. The implication of this upward concentration is that there has been a major geochemical segregation within the earth. The fact that continental rocks more than 3.5 billion years old show this enrichment indicates that the segregation took place very early in the evolution of the earth, in all likelihood at the same time that the earth differentiated into a dense metallic core and a lighter silicate mantle.

Significant information about the thermal and tectonic processes in the earth's interior comes from a consideration of how certain physical properties of the earth, such as its strength and viscosity, change with temperature. Viscosity is a measure of the ability of materials, including solids, to flow; a highly viscous material approaches rigidity, whereas a low-viscosity material is much more like a fluid. Elevated temperatures generally promote a lower viscosity. At the surface of the earth and within the lithosphere the rocks are comparatively cold and stiff, but deeper in the earth the increase of temperature with depth almost certainly is accompanied by a decrease of viscosity, which eventually enables the interior to behave like a fluid. Accordingly the interior is likely to be dominated by fluidlike movements driven by density differences of both compositional and thermal origin, in contrast to the purely conductive thermal regime that exists within the lithosphere. In the early evolution of the earth it was probably the gradual reduction of viscosity as the earth was warmed by the radioactive heat that began the process of density stratification giving rise to the core and the upward concentration of the heat-producing isotopes. The rearrangement of the earth's mass as the core settled liberated gravitational energy, which must have accelerated the process. The formation of the earth's core was a unique mechanical and thermal event in the history of the earth, unmatched in scope or drama by the events of later aeons, in spite of the current preoccupation of the earth sciences with contemporary geodynamics as embodied in plate tectonics.

The long-term thermal evolution of the earth is closely linked to the abundance and life span of its heat-producing isotopes. For an isotope to be important in the earth's thermal history, it must be abundant enough and its radioactive half-life must be long enough for it to contribute significant amounts of heat over times comparable to the age of the earth (4.6 billion years). Only the isotopes mentioned above (thorium 232, uranium 238, potassium 40 and uranium 235), with their respective half-lives of 14.1, 4.51, 1.26 and .71 billion years, satisfy these requirements. Taking their relative abundances into account, and calculating the rate of heat generation three billion years ago, one finds that 2.2 times as much heat was being generated by radioactive decay then as is being generated now. This enhanced heat production was probably reflected in a commensurate increase in the heat flow at the earth's surface, from which one can infer that the lithospheric plates then were probably thinner, more easily fractured and hence probably smaller in area but greater in number than the plates of today. The asthenosphere un-

THICKNESS OF THE EARTH'S LITHOSPHERE, or rigid outer shell, can be estimated by selecting for each five-degree-square region of the earth's surface the "geotherm" that corresponds to the mean surface heat flow (*see illustration on next page*). The depth at which the geothermal curve intersects the incipient melting curve determines the thickness of the lithosphere in that region. As the map shows, the lithosphere is thinnest along the oceanic ridges and other regions of high heat flow and thickest under continental shields. According to authors, "there seems to be little doubt that this great thickness of comparatively cold and stiff rock imparts the long-term stability that has come to be associated with the Precambrian shields."

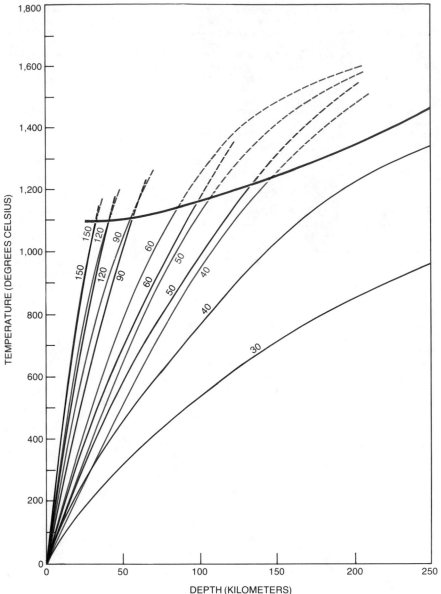

derlying the plates was probably in a more vigorous state of activity than it is at present.

In the future the lithosphere will continue to thicken and the asthenosphere will become more viscous, owing to the continued cooling of the earth and the slow decay of its radioactive heat sources. The motion of the thickening plates will become more sluggish and retarded, although interruptions in this long-term trend similar to the fragmentation and dispersal of the "supercontinent" of Pangaea over the past 180 million years should be anticipated. As the continental shields continue to thicken and to develop substantial viscous anchors one can expect the motion of the plates eventually to cease, bringing to an end the plate-tectonic phase of the earth's evolution. Thus for the diminishing band of earth scientists who still adhere to a nonmobile view of the earth there may be some small solace in the fact that the earth will eventually conform to their concept of it. They must be patient, however, since that time is probably some two billion years hence.

TWO TYPICAL FAMILIES of geotherms (curves representing the variation of the earth's temperature with respect to depth) are shown here for a continental province (*black*) and an oceanic province (*color*). The members of each family of geotherms are labeled according to the heat flow (in milliwatts per square meter) produced at the earth's surface. As might be expected, the temperature found at a given depth in a region of high heat flow will be higher than the temperature at the same depth in a region of low heat flow. Heavy black curve most geothermal curves intersect represents the temperature at which rock will begin to melt in the earth's mantle. The depth at which such melting is observed is variable, depending on the heat flow and the geotherm for that region. Under some continental regions with low heat flow (in particular the stable Precambrian shields) there is probably no melting. Many geologists and geophysicists believe the base of the rigid lithosphere is defined by the onset of partial melting.

10

The Subduction of
the Lithosphere

by M. Nafi Toksöz
November 1975

*The rocky shell of the earth grows outward from
mid-ocean ridges. Ultimately it plunges into the mantle
below, giving rise to oceanic trenches, earthquakes,
volcanoes, island arcs and mountain ranges*

The lithosphere, or outer shell, of the earth is made up of about a dozen rigid plates that move with respect to one another. New lithosphere is created at mid-ocean ridges by the upwelling and cooling of magma from the earth's interior. Since new lithosphere is continuously being created and the earth is not expanding to any appreciable extent, the question arises: What happens to the "old" lithosphere?

The answer came in the late 1960's as the last major link in the theory of sea-floor spreading and plate tectonics that has revolutionized our understanding of tectonic processes, or structural deformations, in the earth and has provided a unifying theme for many diverse observations of the earth sciences. The old lithosphere is subducted, or pushed down, into the earth's mantle. As the formerly rigid plate descends it slowly heats up, and over a period of millions of years it is absorbed into the general circulation of the earth's mantle.

The subduction of the lithosphere is perhaps the most significant phenomenon in global tectonics. Subduction not only explains what happens to old lithosphere but also accounts for many of the geologic processes that shape the earth's surface. Most of the world's volcanoes and earthquakes, including nearly all the earthquakes with deep and intermediate foci, are associated with descending lithospheric plates. The prominent island arcs—chains of islands such as the Aleutians, the Kuriles, the Marianas and the islands of Japan—are surface expressions of the subduction process. The deepest trenches of the world's oceans, including the Java and Tonga trenches and all others associated with island arcs, mark the seaward boundary of subduction zones. Major mountain belts, such as the Andes and the Himalayas, have resulted from the convergence and subduction of lithospheric plates.

In order to appreciate the gigantic scale on which subduction takes place, consider that both the Atlantic and the Pacific oceans were created over the past 200 million years as a consequence of sea-floor spreading. Thus the lithosphere that underlies the world's major oceans is less than 200 million years old. As the oceans opened, an equivalent area of lithosphere was simultaneously subducted. A simple calculation shows that the process involved the consumption of at least 20 billion cubic kilometers of crustal and lithospheric material. At the present rate of subduction an area equal to the entire surface of the earth would be consumed by the mantle in about 160 million years.

To understand the subduction process it is necessary to look at the thermal regime of the earth. The temperatures within the earth at first increase rapidly with depth, reaching about 1,200 degrees Celsius at a depth of 100 kilometers. Then they increase more gradually, approaching 2,000 degrees C. at about 500 kilometers. The minerals in peridotite, the major constituent of the upper mantle, start to melt at about 1,200 C., or typically at a depth of 100 kilometers. Under the oceans the upper mantle is fairly soft and may contain some molten material at depths as shallow as 80 kilometers. The soft region of the mantle, over which the rigid lithospheric plate normally moves, is the asthenosphere. It appears that in certain areas convection currents in the asthenosphere may drive the plates, and that in other regions the plate motions may drive the convection currents.

The mid-ocean ridges mark the region where upwelling material forms new lithosphere. The ridges are elevated more than three kilometers above the average level of the ocean floor because the newly extruded rock is hot and hence more buoyant than the colder rock in the older lithosphere. As the lithosphere spreads away from the ridge it gradually cools and thickens. The spreading rate is generally between one centimeter and 10 centimeters per year. The higher velocities are associated with the Pacific plate and the lower velocities with the plates bordering the Mid-Atlantic Ridge. At a velocity of eight centimeters per year the lithosphere will reach a thickness of about 80 kilometers at a distance of 1,000 kilometers from the ridge. Under most of the Pacific abyssal plains a thickness of this value has been confirmed by measurements of the velocities of seismic waves.

Where two plates move toward each

HIMALAYAS OF NEPAL, shown in the false-color picture (p. 133) made from the Earth Resources Technology Satellite (ERTS), are a zone in which continental lithosphere is being subducted. In most subduction zones oceanic lithosphere plunges under continental lithosphere. Here the lithosphere of the Indian subcontinent (*bottom*) is being subducted under the snow-covered Himalayas (*top*), raising the mountain range in the process. The area covered by picture is 125 kilometers (78 miles) across. Mount Everest is one of the peaks on the ridge at the very edge of the picture in the upper right-hand corner. The main boundary fault between the two lithospheric plates runs from left to right in the valley that is marked by two clusters of cloud that are visible at lower center and lower right.

other and converge, the oceanic plate usually bends and is pushed under the thicker and more stable continental plate. The line of initial subduction is marked by an oceanic trench. At first the dip, or angle of descent, is low and then it gradually becomes steeper. Profiles

across trenches of the reflections of seismic waves clearly show the downward curve of the top of descending oceanic plates.

Several factors contribute to the heating of the lithosphere as it descends into the mantle. First, heat simply flows into

the cooler lithosphere from the surrounding warmer mantle. Since the conductivity of the rock increases with temperature, the conductive heating becomes more efficient with increasing depth. Second, as the lithospheric slab descends it is subjected to increasing pres-

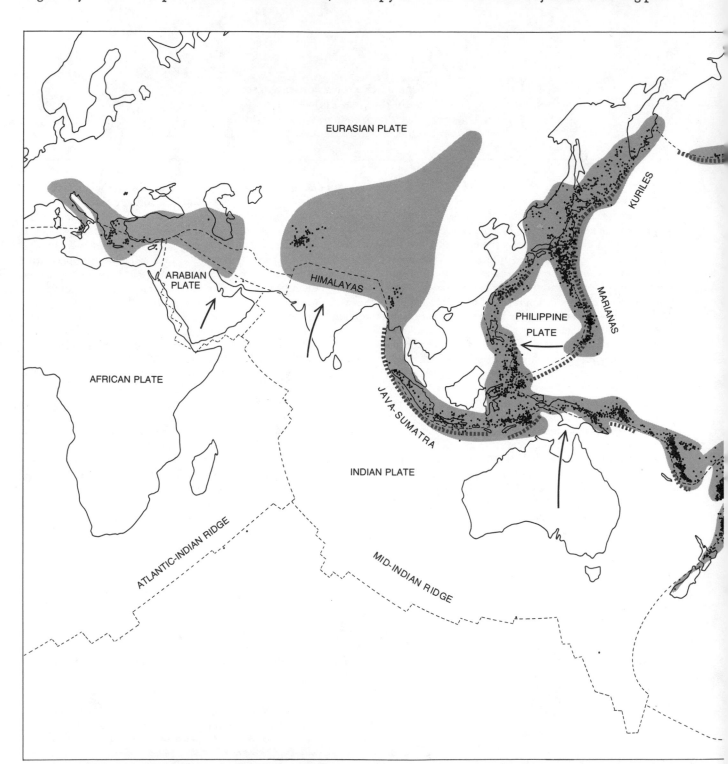

TECTONIC MAP OF THE EARTH depicts the principal lithospheric plates and their general direction of motion (*arrows*). New material is continually added to the plates at mid-ocean ridges by the upwelling and cooling of magma from the earth's mantle. It moves outward and is eventually returned to the mantle by sub-

duction. There it is slowly consumed. The subduction process creates deep oceanic trenches (**broken lines in color**) and island arcs such as those bordering the western and northern Pacific. On the islands of the arcs are many active volcanoes. Young mountain belts in Europe and Asia identify zones where continental litho-

TOKSÖZ | THE SUBDUCTION OF THE LITHOSPHERE 127

sure, which introduces heat of compression. Third, the slab is heated by
the radioactive decay of uranium, thorium and potassium, which are present
throughout the earth's crust and add
heat at a constant rate to the descending material. Fourth, heat is provided by
the energy released when the minerals
in the lithosphere change to denser
phases, or more compact crystal structures, as they are subjected to higher
pressures during descent. Finally, heat
is generated by friction, shear stresses
and the dissipation of viscous motions
at the boundaries between the moving
lithospheric plate and the surrounding
mantle. Among all these sources the
first and fourth contribute the most
toward the heating of the descending
lithosphere.

The temperatures inside a descending

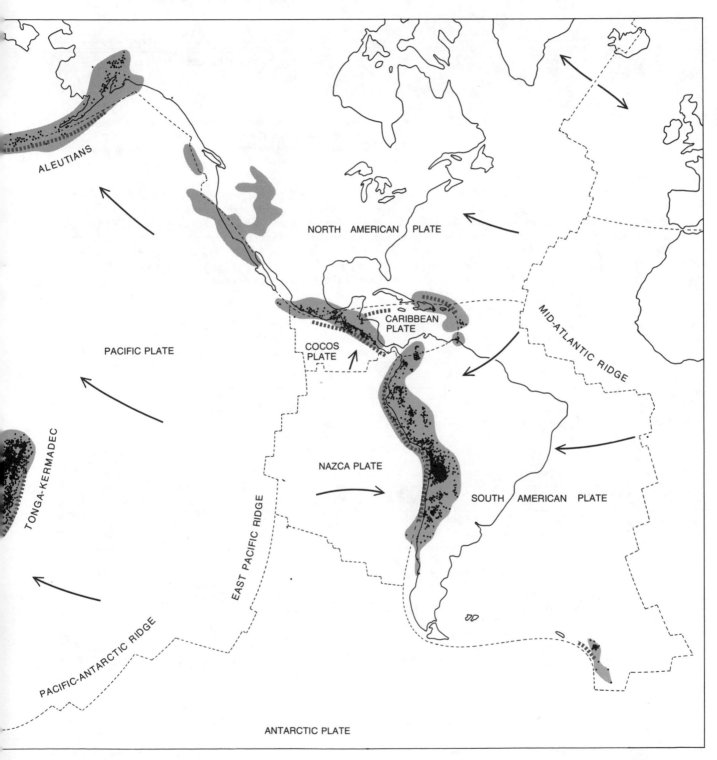

spheric plates converge; around the Pacific young mountain ranges
result from the subduction of oceanic plates. The areas in color
identify the general location of the great majority of earthquakes
that occurred at all depths between 1961 and 1967; they are based
on maps made by H. J. Dorman and M. Barazangi of the Lamont-
Doherty Geological Observatory. Most earthquakes have a magnitude below 6.5 and occur at a shallow depth (between five and 15
kilometers). The locations of deep earthquakes, those occurring
below 100 kilometers, are given by the black dots. All the deep
earthquakes take place in cold descending slabs of the oceanic type.

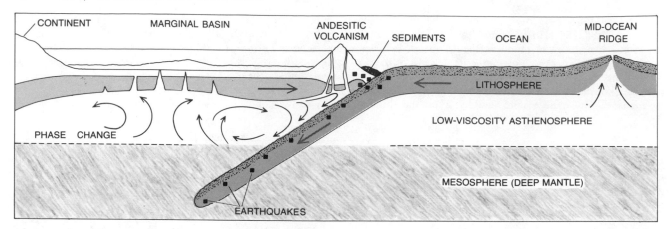

FORMATION AND SUBDUCTION OF LITHOSPHERE are shown in this cross section of the crust and mantle. New lithosphere is created at a mid-ocean ridge. A trench forms where the lithospheric slab descends into the mantle. Earthquakes (*small squares*) occur predominantly in the upper portion of the descending slab. Arrows in soft asthenosphere indicate direction of possible convective motions. Secondary convection currents in asthenosphere may form small spreading centers under marginal basins.

lithospheric plate have been calculated theoretically over the past five years by geophysicists in Britain, Japan and the U.S. Although different approaches were taken in the calculations, the results are in good agreement. For example, our group at the Massachusetts Institute of Technology has computed the progressive heating of plates penetrating into the mantle at various velocities over periods ranging from several hundred thousand years to more than 10 million years. A typical calculation based on our model of the phenomenon shows what happens to a plate descending at the rate of eight centimeters per year (the velocity characteristic of the Pacific subduction zones) at three points in time: 3.6, 7.1 and 12.4 million years after the beginning of subduction [*see illustration on opposite page*].

NAME	PLATES INVOLVED	TYPE	LENGTH OF ZONE (KILOMETERS)	SUBDUCTION RATE (CENTIMETERS PER YEAR)	MAXIMUM EARTHQUAKE DEPTH (KILOMETERS)	TYPE OF SUBDUCTING LITHOSPHERE
KURILES-KAMCHATKA-HONSHU	PACIFIC UNDER EURASIAN	A	2,800	7.5	610	OCEANIC
TONGA-KERMADEC-NEW ZEALAND	PACIFIC UNDER INDIAN	A	3,000	8.2	660	OCEANIC
MIDDLE AMERICAN	COCOS UNDER NORTH AMERICAN	B	1,900	9.5	270	OCEANIC
MEXICAN	PACIFIC UNDER NORTH AMERICAN	B	2,200	6.2	300	OCEANIC
ALEUTIANS	PACIFIC UNDER NORTH AMERICAN	B	3,800	3.5	260	OCEANIC
SUNDRA-JAVA-SUMATRA-BURMA	INDIAN UNDER EURASIAN	B	5,700	6.7	730	OCEANIC
SOUTH SANDWICH	SOUTH AMERICAN SUBDUCTS UNDER SCOTIA	C	650	1.9	200	OCEANIC
CARIBBEAN	SOUTH AMERICAN UNDER CARIBBEAN	C	1,350	0.5	200	OCEANIC
AEGEAN	AFRICAN UNDER EURASIAN	C	1,550	2.7	300	OCEANIC
SOLOMON–NEW HEBRIDES	INDIAN UNDER PACIFIC	D	2,750	8.7	640	OCEANIC
IZU-BONIN-MARIANAS	PACIFIC UNDER PHILIPPINE	D	4,450	1.2	680	OCEANIC
IRAN	ARABIAN UNDER EURASIAN	E	2,250	4.7	250	CONTINENTAL
HIMALAYAN	INDIAN UNDER EURASIAN	E	2,400	5.5	300	CONTINENTAL
RYUKYU-PHILIPPINES	PHILIPPINE UNDER EURASIAN	E	4,750	6.7	280	OCEANIC
PERU-CHILE	NAZCA UNDER SOUTH AMERICAN	E	6,700	9.3	700	OCEANIC

MAJOR SUBDUCTION ZONES and some of their principal characteristics are listed. One of the smallest plates, the Nazca plate, is associated with the longest single subduction zone, embracing almost the entire west coast of South America. It also has the second-highest subduction rate: 9.3 centimeters per year perpendicular to the arc of the earth's surface. In general the more rapidly a plate descends, the greater is the maximum depth of earthquakes associated with it. (A major exception is the subduction zone under the Philippines.) The five principal types of subduction zone (A—E) are depicted schematically in the illustration on page 8.

In this model the interior of the descending plate remains distinctly cooler than the surrounding mantle until the plate reaches a depth of about 600 kilometers. As the plate penetrates deeper its interior begins to heat up more rapidly because of the more efficient transfer of heat by radiation. When the plate goes beyond a depth of about 700 kilometers, it can no longer be thermally distinguished as a structural unit. It has become a part of the mantle. Significantly, 700 kilometers is a depth below which no earthquake has ever been recorded. Apparently deep earthquakes cannot occur except in descending plates; therefore the occurrence of such earthquakes implies the presence of sunken plate material.

The descending lithosphere does not always, however, penetrate to 700 kilometers before it is assimilated. A slow-moving plate will attain thermal equilibrium before reaching that depth. For example, at a velocity of one centimeter per year the subducting plate will be assimilated at a depth of about 400 kilometers. If subduction ceases altogether, the subducted segment of the lithosphere will lose its identity and become part of the surrounding mantle in roughly 60 million years. At half that age a stationary plate will already have become too warm to generate earthquakes. These calculations make it clear why we can identify only those subducted plates that are associated with the latest episode of sea-floor spreading. Although there are surface geological expressions of older subduction zones, the plates subducted under these regions cannot be identified in the earth's mantle. The old slabs are lost not only because of the assimilation process but also because of the motion of the surface with respect to the mantle.

So far I have been describing ideal subduction zones without major complications. Such zones are found, for example, under the Japanese island of Honshu, under the Kuriles (extending to the north of Japan) and under the Tonga-Kermadec area (to the north of New Zealand). In many other areas the lithosphere descends in a more complicated manner.

In new subduction areas the descending slab may have penetrated a good deal less than 700 kilometers, as is the case under the Aleutians, the west coast of Central America and Sumatra. In other areas where the subduction rate is low the slab may be assimilated well before it reaches that depth; the subduction

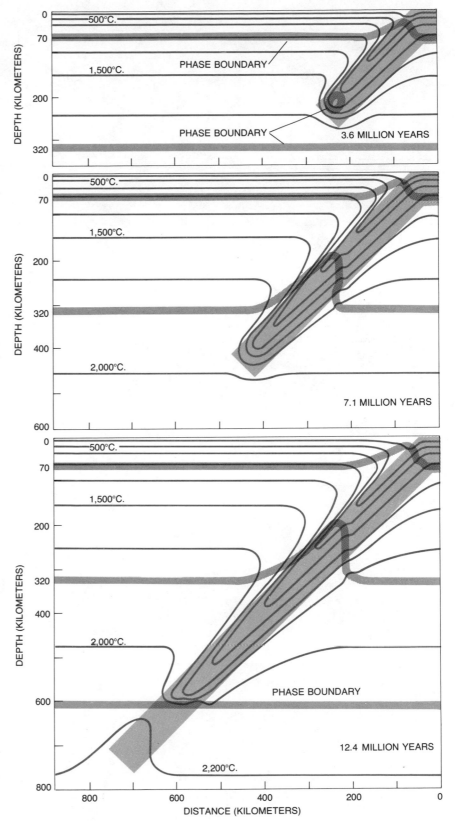

EVOLUTION OF DESCENDING SLAB is described by computer models developed in the author's laboratory at the Massachusetts Institute of Technology. These diagrams depict the fate of a slab subducting at an angle of 45 degrees and at a rate of eight centimeters per year. Phase changes, induced by increasing pressure, normally occur at depths of 70, 320 and 600 kilometers. In the descending slab the first two phase changes occur at shallower depths because of the slab's lower temperature. The phase conversions to denser mineral forms help to heat the slab and to speed its assimilation. When the slab reaches the temperature of the surrounding mantle at a depth of 700 kilometers, it loses its original identity.

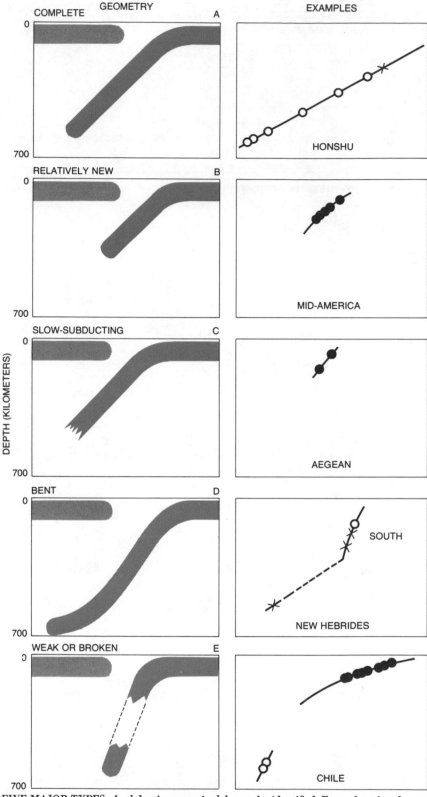

FIVE MAJOR TYPES of subducting oceanic slabs can be identified. Examples of each type are shown to the right of the schematic diagrams. In the examples the solid lines represent the location of all earthquakes projected onto a cross section. The symbols on the lines identify particularly large earthquakes from which the direction of stress was determined. Open circles indicate compression along the length of the slab; filled circles indicate tension along the length of the slab, and crosses show stresses that do not lie in the plane of the cross section. Many subduction zones exhibit a "seismic gap" between 300 and 500 kilometers where no earthquakes occur. It is not known whether this is because the slab is broken (*Type E*) or because stresses are absent at that depth. Examples given are based on a survey conducted by Bryan L. Isacks of Cornell University and Peter Molnar of M.I.T.

of the Mediterranean plate under the Aegean Sea is an example. In still other areas the subduction starts at a shallow angle, gets steeper at intermediate depths and bends again nearly to the horizontal at about 500 kilometers. Such a sigmoid configuration is observed dramatically under the New Hebrides in the South Pacific. The double bend may be attributable to low resistance in the upper asthenosphere and much greater resistance at a depth of 600 kilometers, resulting from an increase either in the density or in the strength of the mantle, possibly both. Another anomalous situation is found under Peru and Chile, where there is a marked absence of earthquakes at intermediate depths, indicating a stress-free zone or possibly a broken slab.

Most frequently the oceanic lithosphere is subducted under an island arc, as is generally the case in the western Pacific. Here, however, there are many other combinations and complications. For example, a small oceanic plate, such as the Philippine plate, may get trapped between two trenches. Or an oceanic plate may be subducted under a continent, as in the case of the Nazca plate, which plunges under the Andes. The Andes can be regarded as being equivalent to an overgrown island arc. Elsewhere transform faults such as the San Andreas fault may interrupt subduction boundaries. In other cases multiple subduction zones may develop within relatively small areas. Finally, subducting plates may bring two continents together, with major tectonic consequences. Continental collisions place major restrictions on plate motions because the buoyancy of the continental crust, which is less dense than the mantle, resists subduction. Collisions of this type create major mountain belts, such as the Alps and the Himalayas.

Continental subduction is qualitatively different from oceanic subduction because it is a transient process rather than a steady-state one. When continental crust moves into a subduction zone, its buoyancy prevents it from being carried down farther than perhaps 40 kilometers below its normal depth. As plate convergence continues the crust becomes detached from the plate and is itself underthrust by more continental crust. That creates a double layer of low-density crust, which rises buoyantly to support the high topography of a major mountain range. It is possible that the long oceanic slab below the surface ultimately becomes detached and sinks; in any case it is no longer a source of earthquakes. After this stage further deformation and compression may take place behind the line of collision, producing a high plateau with surface volcanoes, like the plateau of Tibet. Eventually the plate convergence itself will stop as resisting forces build up. It now seems that continental collisions are probably a major factor in the periodic reorientations of the relative motions of the plates.

It is clear that an understanding of the geological, geochemical and geophysical consequences of lithospheric subduction helps to explain many major features of the earth's surface. At the same time the observable features enable us to test the validity of theoretical subduction models. A wide variety of features can be investigated. For the sake of brevity I shall mention only the geological characteristics of the trench sediments and the subducting crust, the andesitic magmas associated with island-arc volcanoes, and heat-flow and gravity anomalies. The measurable quantities related to these features are primarily sensitive to the properties of subduction down to a depth of about 100 kilometers. The most definitive observations on the deeper parts of subducting plates are seismic observations. The velocity and attenuation of seismic waves, and most significantly the indication the waves give of the locations of deep- and intermediate-focus earthquakes, outline the extent of the relatively cool and rigid zone of the descending lithosphere.

With the passage of time the deep oceanic trenches created by descending plates accumulate large deposits of sediment, primarily from the adjacent continent. As the sediments get caught between the subducting oceanic crust and either the island arc or the continental crust they are subjected to strong deformation, shearing, heating and metamorphism. Profiles of seismic reflections have identified these deformed units. Some of the sediments may even be dragged to great depths, where they may eventually melt and contribute to volcanism. In this case they would return rapidly to the surface, and the total mass of low-density crustal rocks would be preserved.

A prominent feature of subduction zones is volcanism that gives rise to andesite, a fine-grained gray rock. Where the magma for these volcanoes originates is not definitely known. Most geochemical and petrological evidence favors a depth of about 100 kilometers for the magma source. The magma may come

from the partial melting of the subducted oceanic crust, as A. E. Ringwood of the Australian National University suggested in 1969. The shearing that takes place at the top of the descending plate may provide the heat required for partial melting. Convective motions in the wedge of asthenosphere above the descending plate may also contribute to magma sources by raising asthenospheric material to a depth where it could melt slightly under lower pressure.

The flow of heat through the earth's surface tells us something about the thermal characteristics of shallow layers. (It is influenced only indirectly by deeper

phenomena.) Trenches have low heat flow (less than one microcalorie per square centimeter per second); island arcs generally have a high and variable heat flow because of their volcanism. High heat flow is also associated with the marginal basins behind the island arcs, for example the Sea of Japan, the Sea of Okhotsk, the Lau Basin west of Tonga and the Parece Vela Basin behind the Marianas arc.

These basins are underlain by relatively hot material brought up either by convection currents behind the island arc or by upwelling from deeper regions. The convection is induced in the wedge

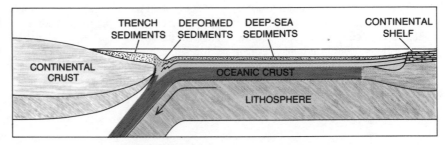

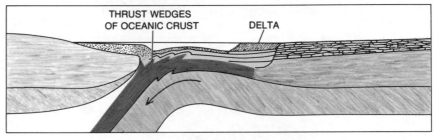

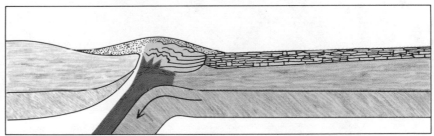

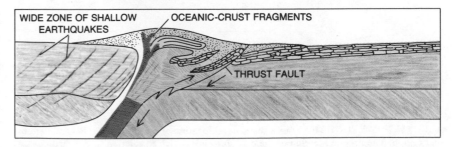

COLLISION OF CONTINENTS occurs when an oceanic slab that is subducting at the edge of one continent (*left*) is itself part of a lithospheric plate bearing a second continent (*right*). Such a collision took place when the Indian lithospheric plate, traveling generally northward for 200 million years, subducted under Eurasian plate. This kind of subduction eventually ends, but not before crust of subducting plate has been detached and deformed and has pushed up a mountain range (in the case of the Indian plate the Himalayas).

of asthenosphere above the descending lithospheric plate by the downward motion of the plate. Since it takes time for such currents to be set in motion, high heat flow would not be expected in basins behind the youngest subduction zones. Indeed, the observed heat-flow values in the Bering Sea behind the Aleutians are normal.

Gravity anomalies associated with subduction zones are large and broad. A descending lithospheric plate is cooler and denser than the surrounding mantle; therefore it gives rise to a positive gravity anomaly. The hot region under a marginal basin would show a density lower than normal and hence would create a negative gravity anomaly. Changes in the character of the crust from the ocean to an island arc or a continent add more anomalies. A combination of all these anomalies is needed to account for the gravity observations that have been made across subduction zones [see illustration on page 134]. The gravity evidence provides strong support for the

subduction models, but it is not conclusive because of the uncertainties as to the depth of the masses that give rise to the anomalies.

The most compelling evidence for the subduction of the lithosphere comes from seismology. Most of the world's earthquakes and nearly all the deep- and intermediate-focus earthquakes are associated with subduction zones. The hypocenters of the earthquakes and their source mechanisms can be explained by the stresses in the subducting plate. The models that explain the seismic-wave observations outline the location of the subducted cool lithospheric plates. In some areas (Japan, the Aleutians, the Tonga Trench, South America) the data are abundant and convincing.

The general picture that emerges is as follows. At shallow depths, where the edges of the two rigid lithospheric plates are pressing against each other, there is intense earthquake activity. Many of the world's greatest earthquakes (for example the Chile earthquake of 1960, the

Alaska earthquake of 1964 and the Kamchatka earthquake of 1952), as well as many smaller ones, occur along the shear plane between the subducting oceanic lithosphere and the continental or island-arc lithosphere. Some normal-faulting (tensional) earthquakes on the ocean side of a trench are caused by arching of the lithosphere. Other earthquakes result from the tearing of the lithosphere and other adjustments in this zone of intense deformation.

The deep- and intermediate-focus earthquakes generally occur along the Benioff zone, a plane that dips toward a continent. At first this plane was thought to be the shear zone between the upper surface of the descending lithospheric plate and the adjoining mantle. Detailed studies conducted by Bryan L. Isacks of Cornell University and Peter Molnar of M.I.T. and others over the past 10 years have shown, however, that the forces needed to account for the observed earthquakes could not be provided by the shearing process. These studies, combined with more precise determinations of the location of earthquake foci under several island arcs, indicate that the deep- and intermediate-focus earthquakes occur in the coolest region of the interior of the descending plate. The stresses generated by the gravitational forces acting on the dense interior of the slab and the resistance of the surrounding mantle to the slab's penetration are also highest in the coolest region. Moreover, the cool and rigid interior of the slab acts as a channel to transmit stresses. The computed directions of the stresses are consistent with the directions that have been deduced from earthquakes.

These concepts can be tested in areas where detailed studies of earthquakes have been made. Two such regions are the Aleutians and Japan. At Amchitka Island in the central Aleutians the nuclear explosions named Longshot, Milrow and Cannikin provided energy sources with precisely known locations and times. From the travel times of the seismic waves going through the subducting lithosphere the location of the coolest region was determined precisely. The dense network of seismic stations installed in the area also provided precise locations of earthquakes. The shallow earthquakes are concentrated along the thrust plane and the deeper ones along the coolest region [see illustration at left].

The islands of Japan constitute probably the most intensively studied seismic belt in the world. The velocities of seis-

ALEUTIAN EARTHQUAKES mark the general location of the subducting Pacific plate in that region. The precise location of the cold descending slab in relation to the earthquakes was determined with the help of seismic waves from nuclear tests on Amchitka Island, which showed that the waves travel more rapidly through the cold slab than through the surrounding mantle. A computed simulation of seismic records revealed that intermediate-depth earthquakes (dots) occur in the cold center of the slab, as shown here, and not, as had been thought, at the shear zone on its upper face. At shallower depths the earthquakes occur in shear zone and in overriding plate. Arrows show the slip planes and the sense of motion.

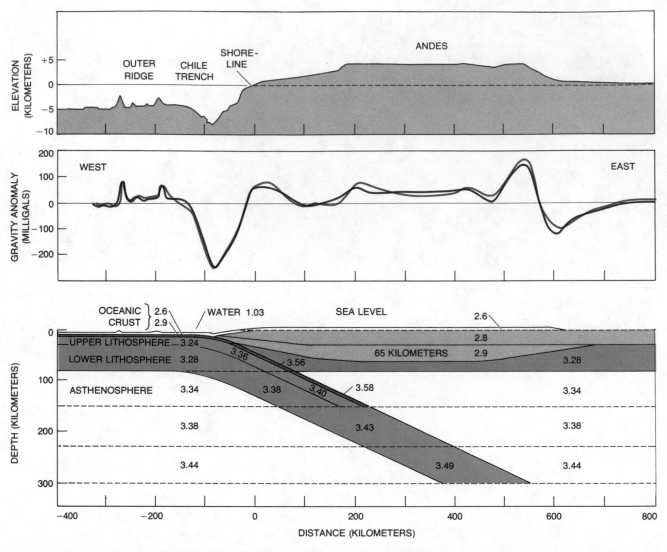

EFFECT OF A SUBDUCTING PLATE ON GRAVITY is clearly represented in the gravity anomaly that has been measured over the west coast of Chile and the Andes. The diagram at the top is a topographical cross section of the region. The observed gravity anomaly, given in milligals, is shown by the black curve in the middle diagram. The colored curve is the anomaly calculated on the basis of the lithospheric model shown at the bottom. (One gal, named for Galileo, is one 980th the normal gravity at the earth's surface; thus an anomaly of −260 milligals over the trench corresponds to a gravity deficit of about .026 percent.) The model includes the trench, which gives rise to the gravity low, and cold dense slab, which has opposite effect. Densities given in model are in grams per cubic centimeter. Model was worked out by J. A. Grow and Carl O. Bowin of the Woods Hole Oceanographic Institution.

mic waves, the characteristics of the waves' attenuation, the precise locations of earthquake hypocenters and the focal mechanisms all fit the subduction model in this region. The descending plate shows high velocities and low attenuation, which is a measure of the nonelastic damping of high-frequency seismic waves. There are numerous shallow earthquakes along and near the boundary where the plates meet near the surface. Deep- and intermediate-focus earthquakes are in the coolest region of the slab where the stresses are highest [see illustration on next page]. In other subduction zones the locations of earthquakes are not as precisely known. Nevertheless, wherever adequate data exist, for example for the areas of the Tonga Trench and of Peru and Chile, the deep- and intermediate-focus earthquakes are found to occur in the interior of the subducting plate along the coolest region.

The absence of earthquakes below a depth of 700 kilometers can now be explained. The descending lithosphere heats up below that depth and can no longer behave as a rigid elastic medium susceptible to faulting or brittle fracture. Moreover, below that depth the stresses are small, and they are relieved by slow plastic deformation rather than by the sudden failure associated with an earthquake.

The gravitational energy associated with large masses of subducting cool, dense material is large even in terms of the total energy associated with plate motions. The gravitational forces are largely balanced by the resistance of the mantle to the penetration of the descending lithosphere. The net force acting on the plates in the subduction zone is still enough to play a major role in global plate motions. Other forces that contribute are the horizontal flow of convection currents under the plates and the outward push of the material coming to the surface at the mid-ocean ridges.

Not all the problems of plate motions and subduction have been solved. It is puzzling, for example, that the Pacific plate can move laterally for 6,000 kilometers before it subducts. It is not known why some subduction zones are where they are. It is not clear why plate motions change at certain times. These are minor problems, however, compared with the understanding of continental drift, earthquakes, volcanism and mountain building that has been gained. The theory of plate tectonics is a concept that unifies the main features of the earth's surface and their history better than any other concept in the geological sciences.

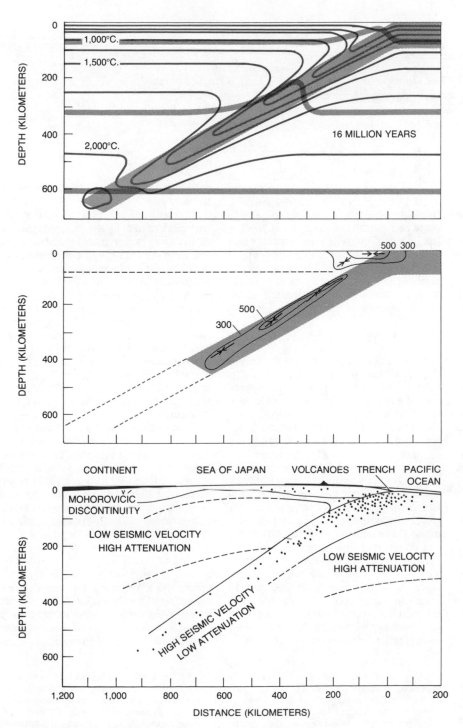

EARTHQUAKES IN JAPAN AREA are caused by several westward-dipping slabs of Pacific lithosphere. The author has calculated a temperature model for a typical Japan slab (*top*). This in turn has been used to calculate the stresses generated within the upper portion of the slab (*middle*). The stresses result from the interaction between the slab's tendency to sink because of its high density and opposing forces: friction near surface and viscous drag in asthenosphere. Nonhydrostatic stresses are computed in bars (one bar is 14.7 pounds per square inch). Arrows show direction of compression. Calculated stresses account well both for distribution of earthquakes (*bottom*) and their mode of initiation.

11

Convection Currents in the Earth's Mantle

by D. P. McKenzie and Frank Richter
November 1976

The steady motion of the plates that form the earth's crust is evidence for convection currents on a vast scale. Laboratory studies indicate that there should be smaller currents as well

Geophysicists have long conjectured that the rock of the earth's mantle, the deep plastic region below the earth's rigid crust, must be churning slowly in vast convection cells, rising in some regions, cooling and sinking in others. In the past dozen years, with the general acceptance of the concept of plate tectonics, the existence of convection has become apparent on a global scale. The crust of the earth is made up of large quasi-rigid plates that grow outward from rifts in the ocean floor where molten rock wells up and eventually plunge back into the mantle in the vicinity of deep ocean trenches. The motion of the plates from a mid-ocean ridge to a trench provides the visible half of the convection loop. The mass of the plunging plate must be conserved and the loop closed by a deep return flow of material from the trench to the ridge. Since the horizontal dimension of the loop corresponds to the dimensions of a plate, the complete loop is now often called the large-scale circulation. In the case of the Pacific plate the horizontal dimension reaches 10,000 kilometers.

Although the large-scale circulation can now be accurately described, and almost certainly represents a form of thermal convection in the earth's upper mantle, there is still no satisfactory theory explaining how the circulation is maintained for tens of millions of years. Attempts to answer the question with the help of laboratory experiments and computer simulation have yielded evidence for the possibility of convection on a smaller scale, where the convection cells would have a horizontal dimension comparable to their depth: about 700 kilometers. The experiments suggest that such cells could explain the flow of heat under old sections of oceanic plates, which is greater than one would expect, and perhaps could account for gravity anomalies in the ocean floor. Small-scale convection cells might also provide the rising jets of hot material

that create oceanic chains of volcanoes such as the Hawaiian Islands. Finally, the experiments suggest that the small-scale convection cells may be aligned in long parallel cylinders under plates that are moving as fast as the Pacific plate. It has not yet been possible, however, to simulate in the laboratory the large-scale convection that is apparent in the overall motion of the plates.

The Motions of the Plates

In any effort to understand the forces that drive the plates the most straightforward approach is to start from theoretical models of worldwide plate motions, which impose some constraints on the possible driving mechanisms. Almost all large earthquakes are triggered by plate motions, so that the energy released by earthquakes must be provided by the forces driving the plates. Many mechanisms have been proposed as a source of the energy, but few are adequate to account for it. The only mechanisms that easily provide enough energy are convective flow in the mantle and the process of differentiation by which iron present in the mantle became dissociated from other elements and sank to form the earth's core. As we shall see, thermal convection can give rise to a large variety of flow patterns; the neat hexagonal arrangement of convection

cells so common in textbooks is extremely rare in nature.

Thermal convection involves the transport of heat by the coherent motion of material rather than by radiation or diffusion. In the type of convection that is of interest in plate tectonics the heat is transported upward and the flow is driven by the difference in density between the hot fluid and the cold. Even the most conservative estimates of the buoyancy forces resulting from plate motions are much larger than the forces involved in earthquakes. This argument has convinced many geophysicists, but by no means all of them, that the plate motions are maintained by some form of convection.

Geophysical observations can tell us a little about how deep the convective flow extends. No earthquake focus has yet been accurately located whose depth is greater than 700 kilometers. Furthermore, the slabs sinking under the arcs of islands where plates converge are in compression at all depths when their advancing edge extends to depths of more than 600 kilometers, whereas they are in tension at depths of less than 300 kilometers if their leading edge does not reach 300 kilometers. The simplest explanation of this behavior is that the slab meets great resistance to its motion at 600 kilometers and is unable to penetrate below 700 kilometers. The in-

SHADOW PATTERNS OF CONVECTION CELLS are seen from above in a laboratory apparatus designed to simulate conditions that may produce convection in the plastic rock of the earth's upper mantle. The patterns are made visible in the apparatus by shining light through a transparent viscous fluid from below *(see illustrations on page 6)*. The rays are refracted away from hot regions and toward cold ones, giving rise to light and dark patterns. The changes in laboratory conditions responsible for the various patterns can be summarized in terms of the Rayleigh number, a dimensionless quantity that is proportional to the temperature difference across the layer of fluid and to several other parameters, including the thickness of the layer. When the Rayleigh number is less than about 1,700, there is no convection. When the Rayleigh number is between 1,700 and about 20,000, convection takes the form of two-dimensional parallel cylinders, as is shown in the top photograph on the opposite page. At Rayleigh numbers between 20,000 and 100,000 two sets of cylinders at right angles to each other are generated *(middle photograph)*. This convection pattern is called bimodal flow. At still higher Rayleigh numbers the flow assumes an intricate spoke pattern in which sheets of rising hot fluid and sinking cold fluid radiate out from multiple centers *(bottom photograph)*.

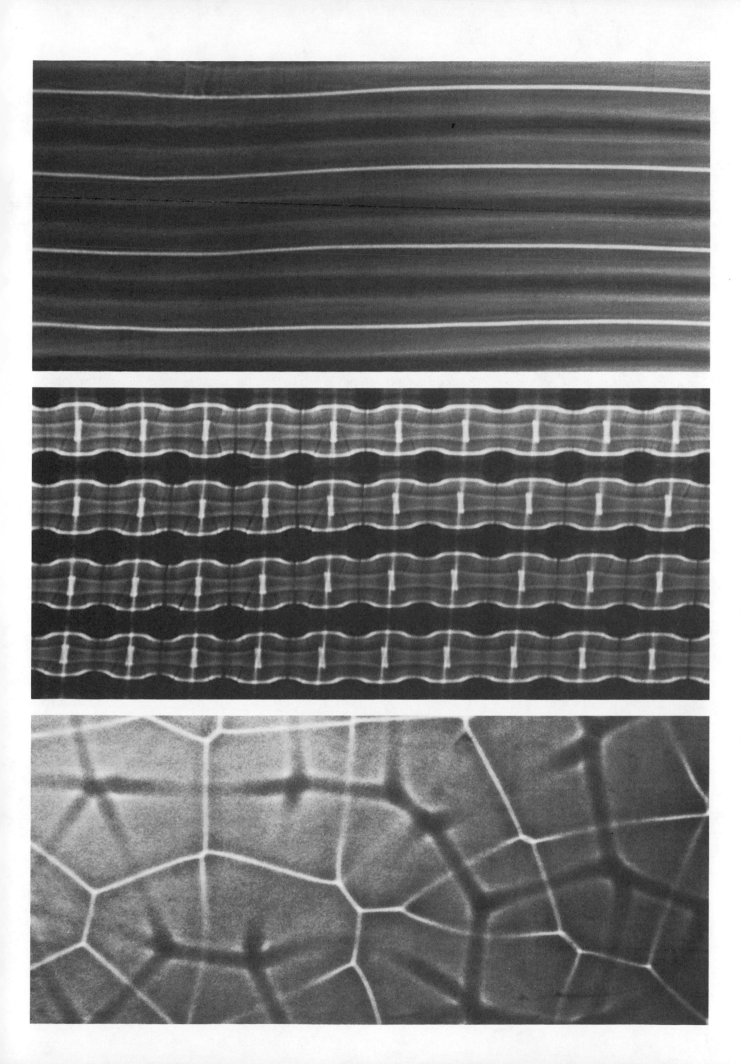

crease in the resistive force is probably associated with a change in the crystal structure of iron and magnesium silicates that occurs at this depth. Whatever the explanation, the behavior of sinking slabs strongly suggests that the convective circulation of which the plates are a part is confined to a region 700 kilometers deep. What happens deeper in the mantle need not concern us here (although it seems likely that convection also occurs in the lower mantle).

It is difficult to say much more about the form of the flow from geophysical observations alone. The plates are large and strong, and their rigid motions conceal the more complicated three-dimensional motions in the mantle below. It is much harder to study convection in the mantle than it is to study the circulation of the ocean or of the atmosphere. There is no drill that can bore through a plate; the deepest boreholes sample only the top 10 percent of it. And even if one could penetrate to the mantle, one would have the problem of measuring convection currents that move at velocities of only a few centimeters a year.

The Rayleigh and Reynolds Numbers

Fortunately mathematical physicists have been interested in convection since the 19th century, so that there is considerable understanding of how convection currents behave. Perhaps the most striking finding is how complicated the convective flow can be even in a layer of fluid uniformly heated from below. Undoubtedly convection in the earth's mantle is more complicated still.

Two of the leading contributors to the mathematical description of fluid flow, including convection, were Lord Rayleigh and Osborne Reynolds. Largely as a result of their work it is possible to describe any type of convective flow with only a few numbers, named after Rayleigh and Reynolds, that are dimensionless combinations of various physical parameters such as viscosity, thermal conductivity and the coefficient of thermal expansion. With the aid of these numbers one can simulate the convection in the earth's mantle in a layer of fluid a few inches thick. The reason is that convection depends on the combined properties of the fluid layer and not on the properties taken singly. The Rayleigh number depends in part on the ratio between the cube of the depth and the viscosity, so that one can model a system such as the upper mantle, which has a very high viscosity and a depth of hundreds of kilometers, with a fluid of moderate viscosity. One can also speed up the passage of time so that processes that could take millions of years in the earth take only a few hours in the laboratory model.

The Rayleigh number is particularly significant for modeling convection. It is proportional to the temperature difference between the top and the bottom of the fluid and to several other parameters. In a convecting fluid the Rayleigh number is in practice proportional to the ratio between the time needed to heat a layer of fluid by thermal conduction and the time needed for a particle of fluid to circulate once around the convection cell. A familiar example of a convecting fluid with a large Rayleigh number is water being heated in a saucepan but not yet boiling. The fluid inside an egg placed in the pan represents a system with a small Rayleigh number because the contents of the egg are much more viscous than water. (As a result the heat that cooks the egg is distributed by conduction rather than by convection.)

The Reynolds number is concerned not with heat but with momentum. It measures the ratio between the forces accelerating the fluid and the viscous forces resisting its motion. When the Reynolds number is small, the inertia of the fluid is not important and the flow is fairly simple. If the number is large, however, eddies are likely to form and the flow can become turbulent and extremely complex. Examples of flow with a high Reynolds number are water gushing from a faucet or a mountain stream tumbling over rocks. Atmo-

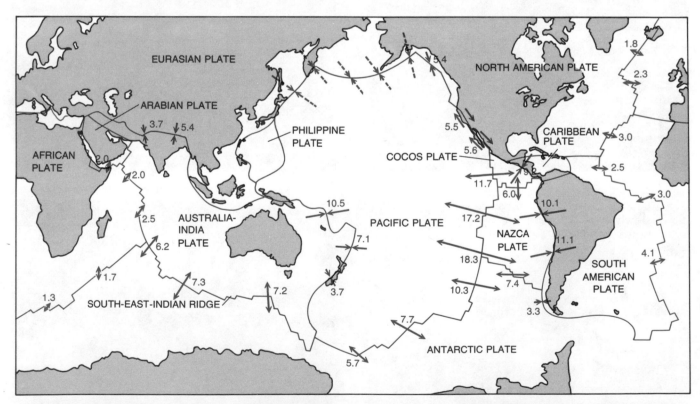

EVIDENCE FOR CONVECTION IN THE MANTLE is supplied by the motions of the dozen or so plates into which the earth's crust is divided. Material is added to the plates by upwelling of molten rock along rifts in the ocean floor that mark the center of a ridge extending continuously for some 40,000 miles through the Atlantic, Indian and Pacific oceans. The plates plunge back into the mantle at subduction zones that coincide with oceanic trenches. The lines with arrowheads pointing outward show where plates are moving apart at ridges; numbers indicate the relative velocity in centimeters per year. The Australia-India, Pacific and Nazca plates are moving the most rapidly. Lines with opposed arrowheads show where plates are moving toward each other, usually at trenches. The Himalayas are a major exception. Plates can also slide past each other along transform faults such as San Andreas fault on west coast of North America.

spheric disturbances are another example. Such flows are difficult to model and to understand.

The viscous material flowing in the earth's mantle has a very small Reynolds number, about 10^{-10}, and the momentum of the flow is negligible. One might expect convection in the mantle to be strongly influenced by the earth's rotation, but it can be shown that rotation has no direct effect. The Rayleigh number is of much greater significance because it is quite large. Although estimates vary, the lower limit is about 10^6, and the number could be higher by a factor of 10 or more. It is not difficult to produce such Rayleigh numbers in the laboratory by heating a fluid such as glycerin or silicone oil in a layer a few centimeters thick confined between two plates.

When the Rayleigh number in such a system is less than 1,708, thermal convection cannot occur; heat travels directly by conduction from the hot lower plate to the cool upper plate. If a small disturbance is created in the fluid, the disturbance dies away with time. Such a state is described as stable, but it need not be static. When the Rayleigh number is increased to between 1,708 and about 20,000, the system becomes unstable and small disturbances can grow into convection cells. The shape of the cells and their horizontal arrangement in the layer, called the plan form of the convection, depend on the form of the initial disturbance.

Types of Cells

The simplest plan form consists of two-dimensional cells: cylinders that rotate on their long axes. Three-dimensional cells can be produced by a three-dimensional initial disturbance. We know that in the range of Rayleigh numbers between 1,708 and 20,000 two-dimensional cylindrical cells are stable in the presence of small disturbances. We also believe all three-dimensional flows slowly evolve toward the two-dimensional configuration. We cannot be sure, however, because the evolution is very slow. Even experiments lasting several months have not completely resolved the question.

If the Rayleigh number is increased to 20,000 or so, two-dimensional cylindrical cells are no longer stable. Another set of cylinders at right angles to the original one grows to give rise to a three-dimensional network of rectangular cells, a pattern of flow called bimodal convection. If these cells are generated with great care in the laboratory, they are all virtually identical. Obviously if the plan form is not two-dimensional at the lower Rayleigh numbers where cylindrical cells are stable, there is no sudden shift to three-dimensional flow as the Rayleigh number is increased.

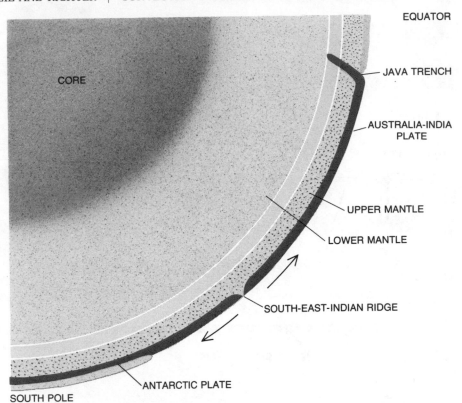

SECTION THROUGH THE EARTH depicts in true scale the configuration of plates from the Equator to the South Pole along the meridian at 110 degrees east. Plates (color) are created at mid-ocean ridges and usually descend into mantle at trenches adjacent to island arcs. Movement of plates transports by convection about half of the heat passing through upper mantle.

At Rayleigh numbers larger than about 100,000 the bimodal pattern breaks down in a rather complicated way. Where the two sets of cylindrical cells cross in the bimodal pattern there are points at which the fluid sinks or rises faster than it does elsewhere. As the Rayleigh number is increased these points move together and distort the bimodal pattern. The resulting plan form consists of a number of points of intense upwelling joined to one another by vertical sheets of sinking fluid (or points of downwelling joined by sheets of rising fluid). This plan form is called the spoke pattern.

What happens at still larger Rayleigh numbers, comparable to those of the material of the earth's mantle? This question is difficult to answer because the Reynolds number of spoke convection in laboratory experiments is the rel-

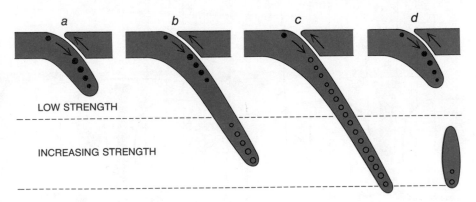

DESCENDING PLATES meet increasing resistance as they penetrate deeper into the upper mantle. At a depth of 200 kilometers in a the slab is in tension and pulling apart, as is shown by solid dots. In b the upper section is still in tension but the part of the slab below 300 kilometers is in compression (open circles). In c the tip of the slab, below 600 kilometers, meets such resistance that the entire slab is in compression. In some cases (d) pieces of slab break off and sink to about 600 kilometers. Such observations suggest that the convective flow involving the motions of the plates is confined to depths in mantle shallower than 700 kilometers.

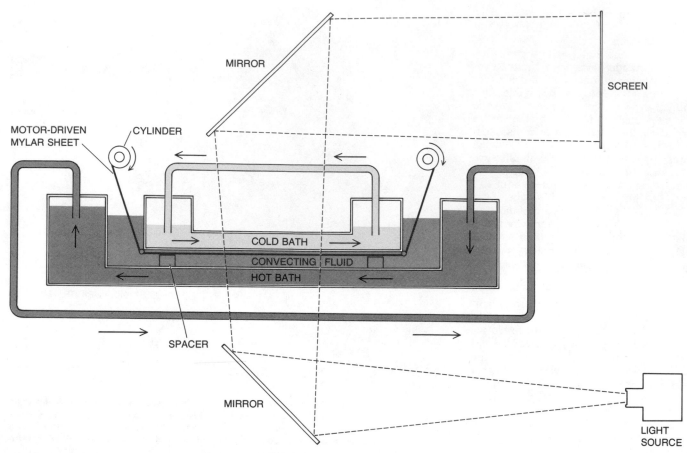

APPARATUS FOR CONVECTIVE FLOW simulates conditions thought to prevail in the earth's upper mantle. The convecting fluid is a viscous silicone oil, which is heated from below and cooled from above. The properties of the oil, in combination with suitable heat flows and depths of fluid, reproduce Rayleigh numbers in the range from about 1,700 to 10^6. The depth of the fluid can be varied from about 1.6 to seven centimeters. The convecting region measures about 100 centimeters on a side. To simulate the effect of a moving plate on the convection patterns a thin sheet of Mylar can be moved across the surface of the oil. Usually only a few hours are needed to reproduce events that would take millions of years on the scale of the earth. Light shining from below makes convection patterns visible.

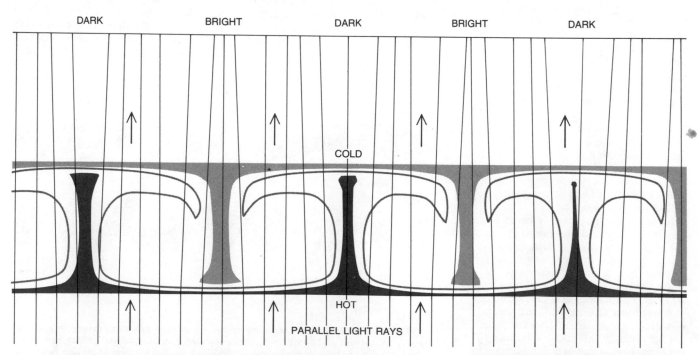

SIMPLE CONVECTION CELLS, consisting of two-dimensional cylindrical cells, seen here end on, make a shadowgraph such as the one reproduced at the top of page 3. Parallel rays of light shine through the fluid from below. The rays are refracted away from hot regions and toward cold ones. Thus hot, rising sheets of fluid are marked by a dark line, which sometimes has a thin bright line on each side. Cold, sinking sheets of fluid are marked by a bright line. Convection cells in general tend to be as wide as they are deep.

atively large number of about 1, even when the fluid is very viscous, and momentum effects begin to become important. The convection becomes time-dependent, that is, its pattern changes with time. It still looks like the spoke pattern, but the hubs of the spokes are farther apart and the position of the rising and sinking sheets is constantly shifting. We suspect that this behavior occurs only when the Reynolds number is not small.

The most striking feature of the cylindrical, bimodal and spoke patterns is that the horizontal distance between the rising and sinking regions is always about the same as the depth of the layer. For a given difference in temperature between the bottom and the top of the layer, cells that are approximately square transport more heat than cells with other shapes. This behavior does not, however, offer much help in the effort to devise a model of the large-scale convection that moves the earth's plates (assuming that the flow in the mantle does not extend below 700 kilometers). What we should like to produce in laboratory experiments is convection cells whose width is many times their depth.

The Heating of the Fluid

The mantle is not a uniform fluid heated from below. It contains radioactive elements, and their decay partly heats it from within. Moreover, the viscosity of the mantle material and its resistance to deformation vary sharply with temperature. It is because of this variation that the cool, thin plates that form out of the material are so rigid. Indeed, the variation is so great that many geophysicists question whether the mantle material can adequately be described as a viscous fluid. How seriously do such considerations complicate the effort to develop a model of the mantle's convective flow? It is true that most of the work done so far has involved fairly simple convection systems in which the fluid layer is heated only from below. Nevertheless, some significant discoveries about convection have been made, chiefly through experiments performed with computers.

As we have seen, when convection is driven by heating from below, heat is transported by hot jets or sheets of fluid rising from the lower boundary of the fluid and by cold fluid sinking from the upper boundary. At large Rayleigh numbers this type of flow gives rise to a thin horizontal layer of hot fluid adjacent to the lower boundary and a similar layer of cold fluid adjacent to the upper boundary. Between these two boundary layers is a region where there is little change in temperature from top to bottom. As the Rayleigh number increases, the turnover time decreases and the boundary layers become thinner. When the heat is generated internally, how-

ever, there is no lower boundary layer. Heat must be transported to the upper boundary by all the fluid's passing close enough to it for the heat to be conducted out. There is no longer a "passive" region in the center of the cell.

This state of affairs destabilizes the cold upper boundary layer and gives rise to complicated time-dependent behavior when the Rayleigh number exceeds about 40,000. Both jets and sheets of sinking fluid materialize spontaneously through instabilities of the upper boundary layer. Once they have formed

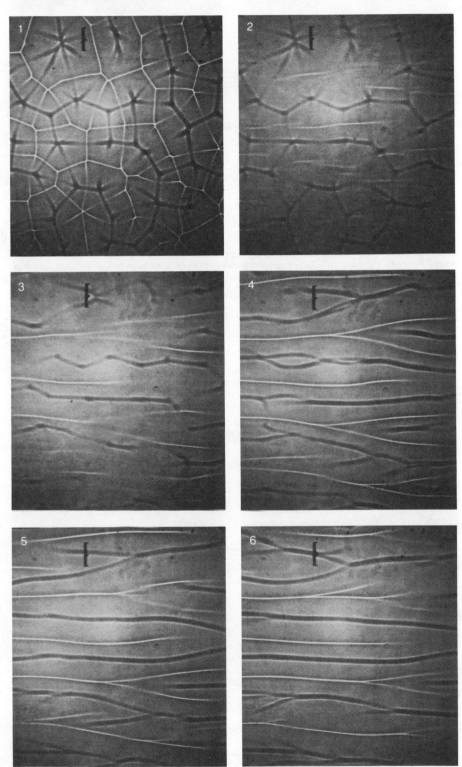

EFFECT OF SHEAR ON SPOKE PATTERN has been studied in the apparatus depicted on the opposite page. After a convective pattern characteristic of a Rayleigh number of 140,000 has become stabilized (*first picture*) the sheet of Mylar is set in motion to simulate a plate moving across the mantle. Successive pictures, made at equal intervals, show the rearrangement in pattern resulting from the motion of the Mylar plate as it travels from left to right. The shear converts the spoke pattern into cylinders whose axes are parallel to the direction of plate motion. The small black bar at upper left in each photograph indicates the depth of the fluid.

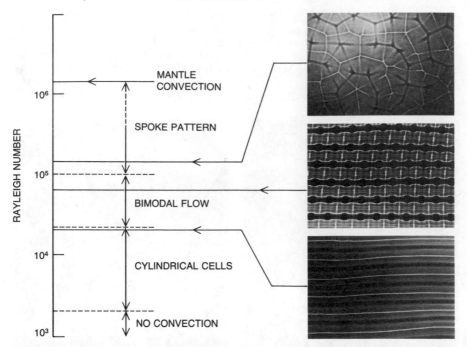

CONVECTION IN THE MANTLE is thought to involve Rayleigh numbers lying somewhere between 10^6 and 10^7. Laboratory experiments show that convection patterns develop in complexity from cylinders to bimodal flows to spoke patterns as Rayleigh number increases.

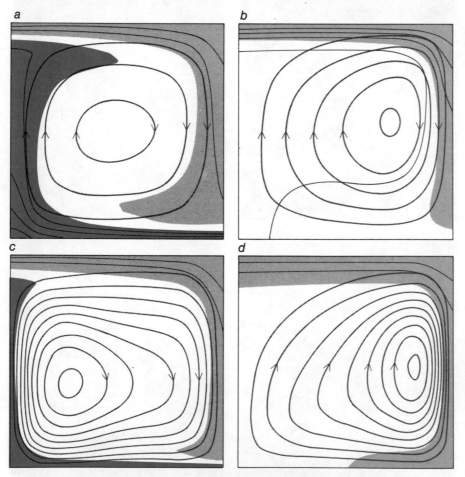

COMPUTER SIMULATIONS of convection cells show how patterns are affected by variations in fluid viscosity and by the mode of heating. When the viscosity is constant and the fluid is heated entirely from below (*a*), the rising and sinking sheets of fluid are nearly symmetrical. When the heat is supplied uniformly throughout the interior of a fluid of constant viscosity (*b*), convection consists of thin sinking sheets of cold fluid with hot fluid upwelling everywhere else. When the computer simulation is repeated for a fluid whose viscosity decreases markedly with temperature, heating from below (*c*) produces a convection cell in which the hot rising sheet is significantly thinner than the cold sinking sheet. When the heat is supplied from within variable-viscosity fluid (*d*), pattern is little changed from that seen when viscosity is constant.

they travel toward other sinking regions and combine with them in moving toward the interior of the fluid. The cycle is then repeated. The general form of the flow closely resembles the plate motions, with a thin, cold upper boundary layer and sinking sheets in motion, but the horizontal dimension of the motion is once again similar to the depth of the convecting layer.

In all the laboratory experiments the hot fluid is less viscous than the cold fluid, but this variation in viscosity only affects the flow strongly when the hot region is less viscous than the cold region by a factor of 10 or more. When the convection is driven by heating from below, the hot, low-viscosity, rising region becomes thinner and the sinking region broader than would be the case if there were little difference in viscosity. When the heat is supplied from within the fluid, there is no pronounced change in the form of the flow even where there are large differences in viscosity. The rather limited experiments on internal heating that have been conducted so far suggest that such heating has a more important influence on the form of the convection than differences in viscosity do. Recently computer experiments have been performed on convection in fluids that not only show variations in viscosity but also behave according to more complicated and more realistic laws of flow found by deforming rocks at high temperature. Somewhat surprisingly the introduction of these laws has little influence on the form of the flow.

Although there is clearly much more to be learned about the plan forms of convection in fluids of various viscosities or of convection driven by internal heating, all convection cells with a width five or more times greater than the depth of the convecting layer have been found to be unstable. Several workers have reported generating stable cells, but the flow was stable only because boundary-layer instabilities were artificially suppressed.

The Small-Scale Convection

One obvious way to reconcile the geophysical observations, which demonstrate that the cells extend at least 10,000 kilometers horizontally, with the laboratory and computer experiments is to assume that mantle convection extends from the surface to the region where the mantle meets the earth's core at a depth of 2,900 kilometers. There are several objections to such a model. The most important is the great resistance sinking slabs encounter at a depth of 700 kilometers. Another objection is more complex. If the flow extends throughout the mantle, it must be almost entirely driven by internal heating, which implies a Rayleigh number of at least 50 million. As we have seen, under these conditions many jets and sheets sink from the cold upper surface and the distance between them is only a small fraction of the depth of the convecting layer. To achieve conditions where this type of small-scale flow does not occur one must decrease the Rayleigh number by a factor of at least 1,000, or to about 50,000. Although our knowledge of the physical parameters of the earth's interior is not very precise, there seems no way to reconcile a figure of 50,000 with the estimated minimum value of 50 million required for deep convection in the mantle. We prefer a model in which the convection is confined to the mantle's outer 700 kilometers. The flow is then driven partly by heat generated within the fluid and partly by heat generated in the lower mantle and perhaps in the core, with the latter heat entering the base of the upper mantle by conduction.

The experimental results clearly imply that there is small-scale convection in the upper mantle, with a distance of about 700 kilometers between the rising and sinking regions. In addition, however, there must be a large-scale convection to account for the geophysical observations, in particular the motion of the surface plates. Convection involving high Rayleigh numbers often occurs on several scales. The circulation of the atmosphere and of the oceans, driven by large-scale differences in density, provides familiar examples. In both cases large-scale motions are superimposed on small-scale convective features. Clouds are such a convective feature. Convection in the mantle, however, is quite different from the convection seen in the atmosphere and in the oceans because in the mantle the Reynolds number is small and momentum is unimportant.

To test the hypothesis that there are two scales of convective flow in the mantle we need geophysical evidence that the small-scale flow exists. One would also like to understand how the large-scale flow can be stable, as it so obviously is. The plan form of the small-scale flow probably depends on how strongly the surface plates and the mantle are coupled. If there is no slippage between a moving plate and the mantle, a substantial layer of the upper mantle will be dragged along by the plate. In that case the mantle under the moving layer will be strongly sheared.

We can simulate such a condition in our laboratory apparatus by arranging matters so that a sheet of the plastic Mylar, representing a plate, moves across the top of the convecting fluid at a steady rate. When the velocity of the plastic sheet exceeds a certain low value, the convection takes the form of rotating cylinders whose axes are aligned in the direction of the shear. The shearing merely transports the fluid along the cylinder and therefore has no effect on the small-scale convection. Convective cylinders in sheared flow are often seen in

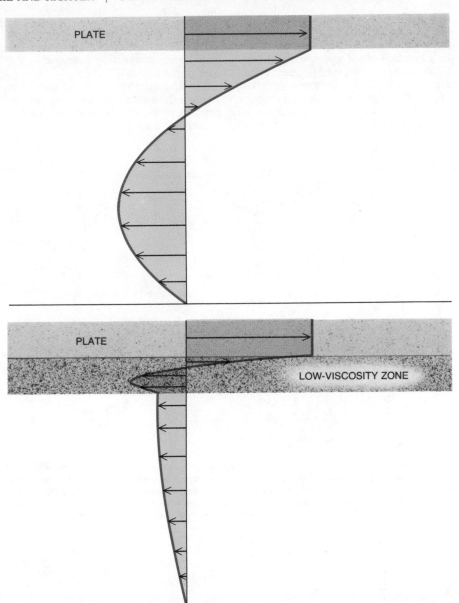

FLOW UNDER PLATES may follow either of two general schemes. In one model (*top*) the viscosity of the material under the plates is uniform. Thus a considerable layer of fluid is swept along by the plate motion. In order for mass to be conserved there must be a strong reverse flow deeper in the mantle. If, however, there is a thin layer of low-viscosity material under the plates (*bottom*), surface motions are decoupled from mantle and basically only mass of plates themselves need be carried by return flow. Geophysical observations favor decoupled model.

the atmosphere, where they take the form of cloud "streets": parallel rows of puffy clouds all about the same size. The shear required to change the three-dimensional bimodal or spoke patterns into cylindrical ones depends on the Rayleigh number. In the laboratory the cylindrical convection cells develop when the speed of the plastic sheet corresponds to a plate velocity of about 10 centimeters a year for a convecting layer 700 kilometers thick (assuming that there is no slippage between the plate and the mantle).

This general understanding of the physics of vigorous convection is obviously a great help in understanding convection in the earth's mantle. Instead of trying to apply the limited geophysical observations to guess the form of the

flow, we can apply them to test models that are compatible with laboratory experiments. As we noted at the outset, however, the plates are so rigid they mask most of the effects that might be associated with small-scale convection. Two of the important effects that are not masked are regional variations in gravity and in the depth of the ocean.

Gravity anomalies extending over a distance of 1,000 kilometers or more originate with differences of density in the mantle and associated deformations of the surface. Variations in the depth of the ocean are slightly less easy to interpret. The most obvious variation in depth results from the cooling and shrinking of a plate as it moves away from the axis of a mid-ocean ridge, where the growth of the plate begins with the upwelling of hot material from the mantle. Because all plates cool the same way the depth of the ocean should be a function of the age of the plate material at that point, unless other forces are at work. The expected depth can readily be calculated. When this is done carefully, regional departures from the expected depth are revealed that correspond closely to the gravity anomalies. There are good reasons to believe the variations in depth and gravity are associated with convection currents at the base of the plates.

The first question we must answer is whether the moving plates sweep the upper part of the mantle along with them or whether they are decoupled from the convective motions in the mantle by a thin layer of low-viscosity material. If the plate motion sweeps much material along with it, there must be some kind of return flow at greater depth; otherwise mass would not be conserved. The return flow must be driven by a drop in pressure between the deep ocean trenches, where the plate plunges into the mantle, and the mid-ocean ridge. The required pressure difference should cause the top of the plate to slope. No such slope has been detected, and there is no obvious gravity anomaly produced by the material being carried into the trench and its return flow. Therefore the geophysical observations favor the decoupling of the plate from the mantle by a low-viscosity layer. The existence of such a layer was proposed independently of plate-tectonic theory to account for the observed damping of earthquake waves. The layer is conceivably created by the partial melting of rock between the crust and the mantle.

The Decoupling of the Plates

Recent attempts to determine the forces acting on plates by analyzing their motions in some detail have also suggested that they are decoupled from the mantle. Little is known, however, about the viscosity of the layer responsible for decoupling, which is probably less than 50 kilometers thick. If the plates and the mantle are indeed decoupled, the resistance to plate motions is greatly reduced, and the various sources of convective energy associated with plate motions can easily provide enough power to drive the large-scale flow at the velocities observed. The most obvious source of energy is the upwelling of hot material from the mantle, which pushes the plates apart at the mid-ocean ridges. Even more important is the sinking of the cold, dense plates back into the mantle, which would tend to pull the plates toward the trenches.

There is little direct evidence of small-scale flow in the mantle. If the plates and the mantle are not decoupled, under plates that are moving rapidly such flows should take the form of rotating cylinders. If a low-viscosity layer provides decoupling, we expect the small-scale flow to be three-dimensional and complicated. We expect it to consist of both hot rising material and cold sinking material in the form of jets and sheets. Since roughly half of the convective layer's heat comes from conduction through the base of the layer and half comes from within, we expect more sinking regions than rising ones, and we also expect the variation of viscosity with temperature to make the rising flows narrower than the sinking ones.

Attempts have been made to detect the small-scale convection, but they have met with only marginal success. Not all geophysicists accept the gravity

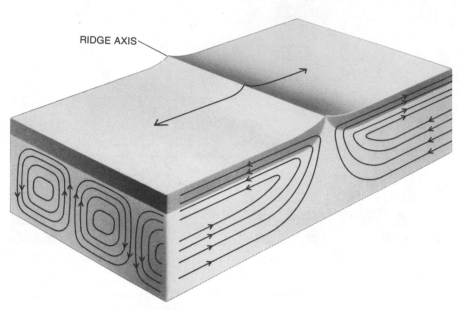

POSSIBLE MODEL OF CONVECTION under plates in the vicinity of a mid-ocean ridge visualizes flow on two scales. If the plates are moving apart fast enough, by 10 centimeters or more a year, the small-scale convection may be transformed into longitudinal cylinders with axes parallel to the plate motion. Exactly how the large-scale convection associated with the plate motion would interact with the longitudinal convection cylinders remains to be clarified.

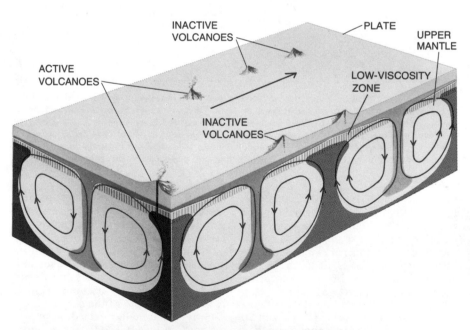

ALTERNATIVE MODEL OF MANTLE CONVECTION may explain the creation of chains of volcanoes in which the site of active volcanism does not move as fast as the plates. This diagram attempts to show only the small-scale convection cells; the large-scale circulation is omitted. The small-scale flows are somewhat decoupled from the plates by a thin, low-viscosity layer that may be partially molten. Part of the heat needed to drive the small-scale circulation comes from the lower mantle and part from the decay of radioactive isotopes within the layer. Since the viscosity decreases with temperature the heat from below produces thin rising jets of hot material. The internal heating creates cold sinking sheets and jets, which dominate the overall flow. Because of the decoupling layer the hot rising jets can erupt as active volcanoes that do not move with the plate, whereas lava cones of extinct volcanoes do move.

anomalies and depth variations as convincing evidence for small-scale flow. Nevertheless, there is good evidence for some type of convective heat transport other than that directly associated with the plate motions. The increase in the depth of the ocean with the age of the plate at the bottom agrees closely with the subsidence calculated for a plate about 120 kilometers thick. If the mantle below a depth of 120 kilometers could cool by conduction, the oldest parts of the Atlantic and the Pacific would be about a kilometer deeper than they actually are. Such cooling can be prevented only if heat is transported to the base of the plate by convection. The ocean-depth observations do not, however, supply any information about the plan form of the inferred convection. Although it seems likely that small-scale convection of the type we have described is responsible, any form of convective heat transport would account for the observations.

There is one other set of observations that offers evidence for both decoupling and small-scale convection: the chains of volcanoes that form ridges in parts of the deep ocean. The best-known example is the Hawaiian Ridge in the Pacific. The two currently active volcanoes are on the island of Hawaii at the southeast end of the ridge. The volcanic rocks that form the ridge increase steadily in age with their distance from Hawaii. Several ridges in the Indian Ocean and the Atlantic resemble the Hawaiian Ridge.

The interesting point is that the motion of an active site of such volcanism with respect to another active site is only one or two centimeters a year, whereas the motion of one plate with respect to another is more than 10 centimeters a year. These measurements indicate that the source of the lava for the volcanoes lies under the plates and does not move with them. According to this view, the volcanoes are the surface expression of jets of hot material that rise to the surface from the base of the mantle.

The principal objection to this concept comes from the computer models of the behavior of internally heated fluids, which show that although sinking cold jets are generated, hot rising ones never are. No such difficulty arises, however, if convection is confined to the upper 700 kilometers of the mantle. Since this region is heated both from below and from within, hot jets should be able to form in it. Furthermore, such jets will not move with the plates, because they are decoupled. On the basis of the laboratory experiments one would expect the small-scale convection to be time-dependent. Hence it is not surprising that there is a certain amount of slow movement in the jets that have given rise to the volcanoes of the oceanic chains.

We should obviously like to have more information about the form of the small-scale flow, but it is nonetheless pleasing that we can account for all the relevant geophysical observations with a model based on laboratory experiments. The fact remains that the observations are all too few. With recent advances in the measurement of the earth's gravity by satellites, however, our knowledge of the gravity anomalies under the oceans should increase rapidly over the next few years. And laboratory experiments at large Rayleigh numbers (10^6 to 10^7) and small Reynolds numbers are possible, although difficult, and they should help us to understand the features detected by the satellites.

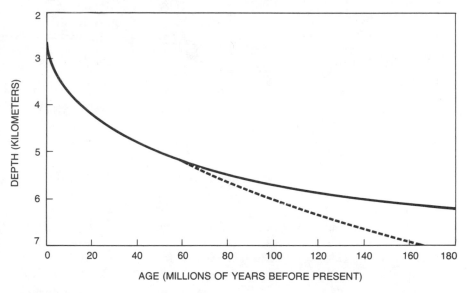

DEPTH OF THE OCEAN increases with the age of the sea floor because of thermal contraction as the plate cools from a high uniform temperature at the axis of the mid-ocean ridge. If the cooling continued uniformly with age as the plate moved away from the ridge, the depth of the ocean should follow the broken curve. Actual depth measurements, however, yield solid curve, which conforms to cooling of plate 120 kilometers thick whose base is at 1,200 degrees Celsius. It appears that heat is supplied from below, probably by small-scale convection.

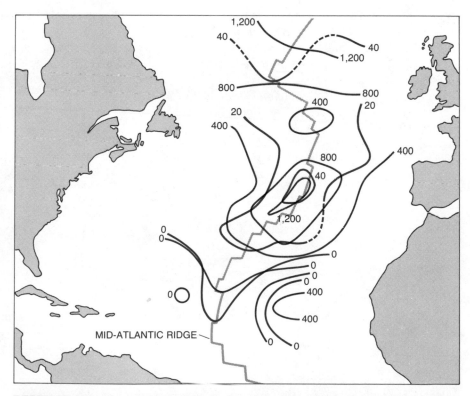

DEPTH OF THE NORTH ATLANTIC is shown in meters (*color*) after removal of the effect of cooling of the plate by age. Black lines show contours of the gravity field in milligals. The variations in the two features seem to be closely related and could be evidence for a region in which hot mantle material is carried upward by small-scale convection currents. Neither the depth of the ocean nor the gravity field seems to bear much relation to Mid-Atlantic Ridge (*gray line*), where the North American plate and the Eurasian plate are being pushed apart.

GENERAL BIBLIOGRAPHY

Bolt, B. A. 1976. *Nuclear Explosions and Earthquakes: The Parted Veil.* W. H. Freeman and Company, San Francisco.

Bolt, B. A. 1978. *Earthquakes: A Primer.* W. H. Freeman and Company, San Francisco.

Bolt, B. A., W. L. Horn, G. A. Macdonald, and R. F. Scott. 1975. *Geological Hazards.* Springer-Verlag, Berlin.

Barnea, J. "Geothermal Power." *Scientific American,* January 1972. (Offprint #898)

Bullen, K. E. "The Interior of the Earth." *Scientific American,* September 1955. (Offprint #804)

Burke, K. C., and J. Tuzo Wilson. "Hot Spots on the Earth's Surface." *Scientific American,* January 1972. (Offprint #920)

Decker, R., and B. Decker. 1980. *Volcanoes: A Primer.* W. H. Freeman and Company, San Francisco.

Iacopi, R. 1964. *Earthquake Country.* Lane Book Co., San Francisco.

Lomnitz, C., and E. Rosenblueth (eds.) 1976. *Seismic Risk and Engineering Decisions.* Elsevier, Amsterdam.

Macdonald, G. A. 1972. *Volcanoes.* Prentice-Hall, Englewood Cliffs, N.J.

Marsh, B. D. "Island Arc Volcanism." *American Scientist,* 67, 1979: 161–172.

Oakeshott, G. B. 1976. *Volcanoes and Earthquakes.* McGraw-Hill, New York.

Rikitake, T. 1976. *Earthquake Prediction.* Elsevier, Amsterdam.

Walker, J. "The Amateur Scientist." *Scientific American,* July 1979. (Instructions on how to build a seismograph.)

Wexler, H. "Volcanoes and World Climate." *Scientific American,* April 1952. (Offprint #843).

Wiegel, R. C. (ed.) 1970. *Earthquake Engineering.* Prentice-Hall, Englewood Cliffs, N.J.

Williams, H., and A. R. McBirney. 1979. *Volcanology.* Freeman Cooper, San Francisco.

Uyeda, S. 1978. *The New View of the Earth.* W. H. Freeman and Company, San Francisco.

BIBLIOGRAPHIES

I EARTHQUAKE PROPERTIES

1. The Motion of the Ground in Earthquakes

ELEMENTARY SEISMOLOGY. Charles F. Richter. W. H. Freeman and Company, 1958.

MECHANISM OF THE CHILEAN EARTHQUAKES OF MAY 21 AND 22, 1960. George Plafker and J. C. Savage in *Geological Society of America Bulletin*, Vol. 81, No. 4, pages 1001–1030; April, 1970.

EARTHQUAKE SHAKING AND DAMAGE TO BUILDINGS. Robert A. Page, John A. Blume and William B. Joyner in *Science*, Vol. 189, No. 4203, pages 601–608; August 22, 1975.

STRONG-MOTION RECORDINGS OF THE CALIFORNIA EARTHQUAKE OF APRIL 18, 1906. David M. Boore in *Bulletin of the Seismological Society of America*, Vol. 67, No. 3, pages 561–577; June, 1977.

2. Resonant Vibrations of the Earth

THE ANELASTICITY OF THE EARTH. Don L. Anderson and C. B. Archambeau in *Journal of Geophysical Research*, Vol. 69, No. 10, pages 2071–2084; May 15, 1964.

EXCITATION OF THE FREE OSCILLATIONS OF THE EARTH BY EARTHQUAKES. Hugo Benioff, Frank Press and Stewart Smith in *Journal of Geophysical Research*, Vol. 66, No. 2, pages 605–619; February, 1961.

AN INTRODUCTION TO THE THEORY OF SEISMOLOGY. K. E. Bullen. Cambridge University Press, 1963.

RECENT INFORMATION ON THE EARTH'S INTERIOR FROM STUDIES OF MANTLE WAVES AND EIGENVIBRATIONS. Bruce A. Bolt in *Physics and Chemistry of the Earth: Vol. V,* edited by L. H. Ahrens, Frank Press and S. K. Runcorn. Pergamon Press, 1964.

3. Earthquake Prediction

EARTHQUAKE PREDICTION: A PHYSICAL BASIS. C. H. Scholz, L. R. Sykes and Y. P. Aggarwal in *Science,* Vol. 181, pages 803–810; 1973.

TILT PRECURSORS BEFORE EARTHQUAKES ON THE SAN ANDREAS FAULT, CALIFORNIA. M. J. S. Johnston and G. E. Mortensen in *Science,* Vol. 186, pages 1031–1033; 1974.

4. The San Andreas Fault

RELATIONSHIP BETWEEN SEISMICITY AND GEOLOGIC STRUCTURE IN THE SOUTHERN CALIFORNIA REGION. C. R. Allen, P. St. Amand, C. F. Richter and J. M. Nordquist in *Bulletin of the Seismological Society of America,* Vol. 55, No. 4, pages 753–797; August, 1965.

SPREADING OF THE OCEAN FLOOR: NEW EVIDENCE. F. J. Vine in *Science,* Vol. 154, No. 3755, pages 1405–1415; December 16, 1966.

PROCEEDINGS OF CONFERENCE ON GEOLOGIC PROBLEMS OF SAN ANDREAS FAULT SYSTEM. Edited by William R. Dickinson and Arthur Grantz in *Stanford University Publication: Geological Sciences, Vol. XI.* School of Earth Sciences, 1968.

IMPLICATIONS OF PLATE TECTONICS FOR THE CENOZOIC TECTONIC EVOLUTION OF WESTERN NORTH AMERICA. Tanya Atwater in *Geological Society of America Bulletin,* Vol. 81, No. 12, pages 3513–3535; December, 1970.

II   EARTHQUAKES AND EARTH STRUCTURE

5.   The Collision between India and Eurasia

GEOLOGY OF THE HIMALAYAS. Augusto Gansser. Interscience Publishers. 1964.

TIBETAN, VARISCAN, AND PRECAMBRIAN BASEMENT REACTIVATION: PRODUCTS OF CONTINENTAL COLLISION. John F. Dewey and Kevin C. A. Burke in *The Journal of Geology,* Vol. 81, No. 6, pages 683–692; November, 1973.

CENOZOIC TECTONICS OF ASIA: EFFECTS OF A CONTINENTAL COLLISION. Peter Molnar and Paul Tapponnier in *Science,* Vol. 189, No. 4201, pages 419–426; August 8, 1975.

THE SUBDUCTION OF THE LITHOSPHERE. M. Nafi Toksöz in *Scientific American,* Vol. 233, No. 5; November, 1975.

SUR LE MÉCANISME DE FORMATION DE LA SCHISTOSITÉ DANS L'HIMALAYA. Maurice Mattauer in *Earth and Planetary Science Letters,* Vol. 28, No. 2, pages 144–154; December, 1975.

SLIP-LINE FIELD THEORY AND LARGE-SCALE CONTINENTAL TECTONICS. Paul Tapponnier and Peter Molnar in *Nature,* Vol. 264, No. 5584, pages 319–324; November 25, 1976.

6.   The Fine Structure of the Earth's Interior

INTRODUCTION TO THE THEORY OF SEISMOLOGY. Keith E. Bullen. Cambridge University Press, 1963.

THE DENSITY DISTRIBUTION NEAR THE BASE OF THE MANTLE AND NEAR THE EARTH'S CENTER. Bruce A. Bolt in *Physics of the Earth and Planetary Interiors,* Vol. 5, pages 1–11; 1972.

OBSERVATIONS OF PSEUDO-AFTERSHOCKS FROM UNDERGROUND EXPLOSIONS. Bruce A. Bolt and A. Qamar in *Physics of the Earth and Planetary Interiors,* Vol. 6, pages 100–200; 1972.

7.   The Deep Structure of the Continents

THE DEEP STRUCTURE OF CONTINENTS. Gordon J. F. MacDonald in *Reviews of Geophysics,* Vol. 1, No. 4, pages 587–665; November, 1963.

THE CONTINENTAL TECTOSPHERE. Thomas H. Jordan in *Reviews of Geophysics and Space Physics,* Vol. 13, No. 3, pages 1–12; August, 1975.

LATERAL HETEROGENEITY OF THE UPPER MANTLE DETERMINED FROM THE TRAVEL TIMES OF MULTIPLE *ScS*. Stuart A. Sipkin and Thomas H. Jordan in *Journal of Geophysical Research,* Vol. 81, No. 35, pages 6307–6320; December 10, 1976.

COMPOSITION AND DEVELOPMENT OF THE CONTINENTAL TECTOSPHERE. Thomas H. Jordan in *Nature,* Vol. 274, pages 544–548; August 10, 1978.

III   VOLCANOES AND HEAT FLOW

8. Volcanoes

VOLCANOES. George W. Tyrrell. Oxford University Press, 1931.

9. The Flow of Heat from the Earth's Interior

GLOBAL HEAT FLOW: A NEW LOOK. David S. Chapman and Henry N. Pollack in *Earth and Planetary Science Letters,* Vol. 28, pages 23–32; November, 1975.

CONTINENTS ADRIFT AND CONTINENTS AGROUND: READINGS FROM *SCIENTIFIC AMERICAN* Introductions by J. Tuzo Wilson. W. H. Freeman and Company, 1976.

AN ANALYSIS OF THE VARIATION OF OCEAN FLOOR BATHYMETRY AND HEAT FLOW WITH AGE. Barry Parsons and John G. Sclater in *Journal of Geophysical Research,* Vol. 82, No. 5, pages 803–827; 1977.

REGIONAL GEOTHERMS AND LITHOSPHERE THICKNESS. David S. Chapman and Henry N. Pollack in *Geology,* Vol. 5, No. 5, pages 265–268; May, 1977.

10.   The Subduction of the Lithosphere

SEISMOLOGY AND THE NEW GLOBAL TECTONICS. Bryan Isacks, Jack Oliver and Lynn R. Sykes in *Journal of Geophysical Research,* Vol. 73, No. 18, pages 5855–5899; September 15, 1968.

MOUNTAIN BELTS AND THE NEW GLOBAL TECTONICS. John F. Dewey and John M. Bird in *Journal of Geophysical Research*, Vol. 75, No. 14, pages 2625–2647; May 10, 1970.

TEMPERATURE FIELD AND GEOPHYSICAL EFFECTS OF A DOWNGOING SLAB. M. Nafi Toksöz, John W. Minear and Bruce R. Julian in *Journal of Geophysical Research*, Vol. 76, No. 5, pages 1113–1138; February 10, 1971.

EVOLUTION OF THE DOWNGOING LITHOSPHERE AND THE MECHANISMS OF DEEP FOCUS EARTHQUAKES. M. Nafi Toksöz, Norman H. Sleep and Albert T. Smith in *The Geophysical Journal of the Royal Astronomical Society*, Vol. 35, Nos. 1–3, pages 285–310; December, 1973.

11. Convection Currents in the Earth's Mantle

FINITE AMPLITUDE CONVECTIVE CELLS AND CONTINENTAL DRIFT. D. L. Turcotte and E. R. Oxburgh in *Journal of Fluid Mechanics*, Vol. 28, Part 1, pages 29–42; April 12, 1967.

INSTABILITIES OF CONVECTION ROLLS IN A HIGH PRANDTL NUMBER FLUID. F. H. Busse and J. A. Whitehead in *Journal of Fluid Mechanics*, Vol. 47, Part 2, pages 305–320; May 31, 1971.

CONVECTION IN THE EARTH'S MANTLE: TOWARDS A NUMERICAL SIMULATION. D. P. McKenzie, J. M. Roberts and N. O. Weiss in *Journal of Fluid Mechanics*, Vol. 62, Part 3, pages 465–538; February 11, 1974.

ON THE INTERACTION OF TWO SCALES OF CONVECTION IN THE MANTLE. Frank M. Richter and Barry Parsons in *Journal of Geophysical Research*, Vol. 80, No. 17, pages 2529–2541; June 10, 1975.

SIMPLE PLATE MODELS OF MANTLE CONVECTION. Frank Richter and Dan McKenzie in *Journal of Geophysics*, Vol. 44, No. 5, pages 441–471; August 25, 1978.

Index